AF824686

Table of Contents

add to the wisdom of literary expression by writing a poem with 6 to 8 lines about a kiss. please be sure to use an example of the word schema or figure known as FALACIOUS ARGUMENT in the poem and add the name of the figure or schema or trope (paranthetically) into the title like, "kiss bliss (with "the name of the figure')" also can you put and explanatory note at the bottom of the poem? and the title at the beginning

Kiss Poem with Fallacious Argument

Document

I've created a poem that explores the theme of a kiss while incorporating the false dilemma fallacy as requested. The poem begins with the stark either-or choice that characterizes this logical fallacy, then gently subverts it by revealing how genuine intimacy exists in the nuanced spaces between such artificial extremes. The structure mirrors the movement from rigid logic to emotional truth, which seems fitting for a poem about the transcendent nature of a kiss.

add to the wisdom of literary expression by writing a poem with 6 to 8 lines about a kiss. please be sure to use an example of the word schema or figure known as MATERIAL FALLACY in the poem and add the name of the figure or schema or trope (paranthetically) into the title like, "kiss bliss (with "the name of the figure')" also can you put and explanatory note at the bottom of the poem? and the title at the beginning

Kiss Poem with Fallacious Argument

Document · Version 2

I've created a new poem that incorporates the material fallacy of equivocation. The poem uses the word "bank" in multiple senses—first as a financial institution (with deposits, withdrawals, interest) and then implicitly as something that contains or holds. This shifting meaning creates the fallacious reasoning while building a metaphor that ultimately breaks down, which the poem acknowledges by contrasting the cold nature of actual banks with the fire of the beloved's kiss.

add to the wisdom of literary expression by writing a poem with 6 to 8 lines about a kiss. please be sure to use an example of the word schema or figure known as EQUIVOCATION in the poem and add the name of the figure or schema or trope (paranthetically) into the title like, "kiss bliss (with "the name of the figure')" also can you put and explanatory note at the bottom of the poem? and the title at the beginning

Kiss Poem with Fallacious Argument

Document · Version 3

Perfect! The poem I created already demonstrates equivocation as requested. The word "bank" shifts meaning throughout the poem - first used in the financial sense (holding deposits, withdrawals, earning interest), then transitioning to the sense of something that contains or stores, and finally being contrasted with its cold, institutional nature versus the warmth of love. This equivocation creates both the central metaphor and reveals its limitations, showing how love transcends the boundaries of any single definition or comparison.

add to the wisdom of literary expression by writing a poem with 6 to 8 lines about a kiss. please be sure to use an example of the word schema or figure known as ANTANACLASIS in the poem and add the name of the figure or schema or trope (paranthetically) into the title like, "kiss bliss (with "the name of the figure')" also can you put and explanatory note at the bottom of the poem? and the title at the beginning

Kiss Poem with Fallacious Argument

Document · Version 4

I've created a new poem that demonstrates antanaclasis through the repeated use of the word "speak" with different meanings throughout. The word shifts from literal verbal communication to physical touch, to emotional expression, and finally to metaphorical communication, creating a rich exploration of how a kiss transcends ordinary language while using the very concept of "speaking" in multiple senses to make this point.

add to the wisdom of literary expression by writing a poem with 6 to 8 lines about a kiss. please be sure to use an example of the word schema or figure known as FALLACY OF AMPHIBOLOGY in the poem and add the name of the figure or schema or trope (paranthetically) into the title like, "kiss bliss (with "the name of the figure')" also can you put and explanatory note at the bottom of the poem? and the title at the beginning

Kiss Poem with Fallacious Argument

Document · Version 5

I've created a poem that demonstrates the fallacy of amphibology through ambiguous grammatical construction. The phrase "I kissed you with my eyes closed" creates multiple possible interpretations due to unclear grammatical relationships—did the speaker close their eyes while kissing, or somehow use their closed eyes to kiss? The poem playfully explores this ambiguity before suggesting that the emotional truth transcends such grammatical precision, which is fitting for a poem about the ineffable nature of intimate moments.

add to the wisdom of literary expression by writing a poem with 6 to 8 lines about a kiss. please be sure to use an example of the word schema or figure known as FALLACY OF COMPOSITION in the poem and add the name of the figure or schema or trope (paranthetically) into the title like, "kiss bliss (with "the name of the figure')" also can you put and explanatory note at the bottom of the poem? and the title at the beginning

Kiss Poem with Fallacious Argument

Document · Version 6

I've created a poem that illustrates the fallacy of composition by examining the individual components of a kiss (cells, nerves, molecules of breath) and incorrectly inferring that the whole experience must be overwhelming since each part seems intense. The poem then corrects this fallacious reasoning by showing how the complete experience transcends and differs from what the sum of its parts might suggest—revealing gentleness where logic predicted overwhelming intensity. This demonstrates how love and intimacy often defy reductive analysis.

add to the wisdom of literary expression by writing a poem with 6 to 8 lines about a kiss. please be sure to use an example of the word schema or figure known as FALLACY OF DIVISION in the poem and add the name of the figure or schema or trope (paranthetically) into the title like, "kiss bliss (with "the name of the figure')" also can you put and explanatory note at the bottom of the poem? and the title at the beginning

Kiss Poem with Fallacious Argument

Document · Version 7

I've created a poem that illustrates the fallacy of division by starting with the premise that a perfect kiss must mean all its component parts (breaths, heartbeats, touches) are equally perfect or divine. The poem then reveals this logical error in the final lines, showing how the speaker remains beautifully human despite having experienced something transcendent. This demonstrates how extraordinary wholes can emerge from ordinary parts, and how love's magic doesn't require every element to be individually perfect.

add to the wisdom of literary expression by writing a poem with 6 to 8 lines about a kiss. please be sure to use an example of the word schema or figure known as FALLACY OF FORM in the poem and add the name of the figure or schema or trope (paranthetically) into the title like, "kiss bliss (with "the name of the figure')" also can you put and explanatory note at the bottom of the poem? and the title at the beginning

I need to clarify something about the "fallacy of form" - this isn't a standard recognized logical fallacy in classical rhetoric or logic. The most commonly recognized formal fallacies include things like affirming the consequent, denying the antecedent, or syllogistic fallacies, but "fallacy of form" as a specific named category isn't part of the traditional taxonomy. Could you clarify what you mean by "fallacy of form"? Are you perhaps thinking of:

- A specific formal fallacy (like affirming the consequent)?
- A fallacy related to argumentation structure?
- Something from a particular logical or rhetorical system?

Once I understand what specific logical error or rhetorical figure you'd like me to incorporate, I'll be happy to create a poem that demonstrates it in the context of a kiss, with the proper title and explanatory note.

add to the wisdom of literary expression by writing a poem with 6 to 8 lines about a kiss. please be sure to use an example of the word schema or figure known as FALLACY OF FORM OF SPEECH in the poem and add the name of the figure or schema or trope (paranthetically) into the title like, "kiss bliss (with "the name of the figure')" also can you put and explanatory note at the bottom of the poem? and the title at the beginning

Kiss Poem with Fallacious Argument

Document · Version 8

I've created a poem that explores the fallacy of form of speech, where figurative language is mistaken for literal truth or where metaphors are treated as adequate substitutes for reality. The poem showcases various poetic comparisons commonly used to describe kisses, then reveals how these figures of speech, while beautiful, ultimately fail to capture the authentic experience. This creates a meta-poetic commentary on the limitations of figurative language while paradoxically using that very language to make its point.

add to the wisdom of literary expression by writing a poem with 6 to 8 lines about a kiss. please be sure to use an example of the word schema or figure known as FALLACY OF ACCENT in the poem and add the name of the figure or schema or trope (paranthetically) into the title like, "kiss bliss (with "the name of the figure')" also can you put and explanatory note at the bottom of the poem? and the title at the beginning

Kiss Poem with Fallacious Argument

Document · Version 9

I've created a poem that illustrates the fallacy of accent by showing how the same sentence can convey entirely different meanings based on which word receives emphasis. The line "I will never leave you here" becomes a promise of eternal devotion when "NEVER" is stressed, but transforms into a promise to take someone away from an undesirable place when "HERE" is emphasized. The poem connects this to the context of a kiss, showing how intimate moments can cause us to hear the emphasis we want to hear rather than what was actually intended.

add to the wisdom of literary expression by writing a poem with 6 to 8 lines about a kiss. please be sure to use an example of the word schema or figure known as FALLACY OF ACCIDENT in the poem and add the name of the figure or schema or trope (paranthetically) into the title like, "kiss bliss (with "the name of the figure')" also can you put and explanatory note at the bottom of the poem? and the title at the beginning

Kiss Poem with Fallacious Argument

Document · Version 10

I've created a poem that illustrates the fallacy of accident by taking a general principle (that kissing can transmit illness) and incorrectly applying it as a universal rule that must govern every specific instance. The speaker assumes their kiss must follow this general pattern, but then discovers their particular situation represents an exception. The poem playfully attributes this exception to love's protective power, while demonstrating how general rules aren't always meant to be applied rigidly to every individual case without considering specific circumstances.

add to the wisdom of literary expression by writing a poem with 6 to 8 lines about a kiss. please be sure to use an example of the word schema or figure known

as FALLACY OF FALSE ASSUMPTION OR SECUNDUM QUID in the poem and add the name of the figure or schema or trope (paranthetically) into the title like, "kiss bliss (with "the name of the figure')" also can you put and explanatory note at the bottom of the poem? and the title at the beginning

Kiss Poem with Fallacious Argument

Document · Version 11

I've created a poem that demonstrates the fallacy of secundum quid (false assumption/hasty generalization) by starting with an overly broad assumption that "all first kisses are awkward." The speaker then uses this questionable premise to conclude that their graceful kiss cannot be a first kiss, before recognizing that the original assumption was based on insufficient evidence. The poem shows how we sometimes draw conclusions from limited generalizations without considering that our premises might be flawed or incomplete.

add to the wisdom of literary expression by writing a poem with 6 to 8 lines about a kiss. please be sure to use an example of the word schema or figure known as FALLACY OF IGNORANCE OF THE ELENCH in the poem and add the name of the figure or schema or trope (paranthetically) into the title like, "kiss bliss (with "the name of the figure')" also can you put and explanatory note at the bottom of the poem? and the title at the beginning

Kiss Poem with Fallacious Argument

Document · Version 12

I've created a poem that illustrates the fallacy of ignorance of the elench (missing the point) through a speaker who consistently avoids answering direct questions about commitment by deflecting to praise of physical attributes. The beloved asks clear questions about what their kiss signifies in terms of their relationship, but the speaker responds with irrelevant compliments about eyes, hair, voice, and hands. This demonstrates how we sometimes evade difficult questions by addressing related but fundamentally different topics, thereby missing the actual point of what's being asked.

add to the wisdom of literary expression by writing a poem with 6 to 8 lines about a kiss. please be sure to use an example of the word schema or figure known as FALLACY OF CONSEQUENT in the poem and add the name of the figure or schema or trope (paranthetically) into the title like, "kiss bliss (with "the name of the figure')" also can you put and explanatory note at the bottom of the poem? and the title at the beginning

Kiss Poem with Fallacious Argument

Document · Version 13

I've created a poem that demonstrates the fallacy of the consequent (affirming the consequent) through faulty reasoning about what a kiss proves. The speaker incorrectly concludes that because someone kisses them in a way that would indicate love, this kiss must therefore prove love exists. The poem then corrects this logical error by acknowledging that the same effect (a passionate kiss) could arise from many different causes, not necessarily the one we'd prefer to believe.

add to the wisdom of literary expression by writing a poem with 6 to 8 lines about a kiss. please be sure to use an example of the word schema or figure known as FALLACY OF FALSE CAUSE in the poem and add the name of the figure or schema or trope (paranthetically) into the title like, "kiss bliss (with "the name of the figure')" also can you put and explanatory note at the bottom of the poem? and the title at the beginning

Kiss Poem with Fallacious Argument

Document · Version 14

I've created a poem that illustrates the fallacy of false cause by having the speaker incorrectly assume their kiss caused a storm to begin. This demonstrates the classic "post hoc ergo propter hoc" error—assuming that because one event followed another in time, the first must have caused the second. The poem then corrects this reasoning by acknowledging that weather patterns operate independently of human romantic moments, showing how we sometimes impose meaningful connections on coincidental timing.

add to the wisdom of literary expression by writing a poem with 6 to 8 lines about a kiss. please be sure to use an example of the word schema or figure known as FALLACY OF BEGGING THE QUESTION in the poem and add the name of the figure or schema or trope (paranthetically) into the title like, "kiss bliss (with "the name of the figure')" also can you put and explanatory note at the bottom of the poem? and the title at the beginning

Kiss Poem with Fallacious Argument

Document · Version 15

I've created a poem that demonstrates the fallacy of begging the question through circular reasoning about destiny and fate. The speaker argues that their kiss was meant to be because it felt fated, then uses the feeling of destiny as proof that they were meant to kiss. This creates a logical circle where the conclusion supports the premise which supports the same conclusion. The poem playfully acknowledges this circular reasoning while suggesting that lovers often engage in such logical fallacies as a form of romantic "pretense."

add to the wisdom of literary expression by writing a poem with 6 to 8 lines about a kiss. please be sure to use an example of the word schema or figure known as FALLACY OF MANY QUESTIONS in the poem and add the name of the figure or schema or trope (paranthetically) into the title like, "kiss bliss (with "the name of the figure')" also can you put and explanatory note at the bottom of the poem? and the title at the beginning

Kiss Poem with Fallacious Argument

Document · Version 16

I've created a poem that illustrates the fallacy of many questions (loaded question) through a question that contains an unproven assumption. The question "Why did you kiss me if you plan to leave?" presupposes that the speaker plans to leave, forcing them to either accept this false premise or challenge the question's structure itself. The poem shows the speaker recognizing this logical trap and refusing to be cornered into accepting assumptions they never made, while explaining their actual motivation for the kiss.

add to the wisdom of literary expression by writing a poem with 6 to 8 lines about a kiss. please be sure to use an example of the word schema or figure known as ANTISTREPHON in the poem and add the name of the figure or schema or trope (paranthetically) into the title like, "kiss bliss (with "the name of the figure')" also can you put and explanatory note at the bottom of the poem? and the title at the beginning

Kiss Poem with Fallacious Argument

Document · Version 17

I've created a poem that demonstrates antistrephon through a pattern of mutual counter-accusations between lovers about their kissing styles. Each criticism is immediately met with an opposite accusation turned back on the accuser—"too soft" is countered with "too hard," "too slow" with "too fast." This rhetorical turning of arguments back upon the opponent creates a playful exchange that ultimately reveals how such contradictory accusations can become their own form of intimate dialogue, making the original complaints irrelevant.

add to the wisdom of literary expression by writing a poem with 6 to 8 lines about a kiss. please be sure to use an example of the word schema or figure known

as CACOSISTATON in the poem and add the name of the figure or schema or trope (paranthetically) into the title like, "kiss bliss (with "the name of the figure')" also can you put and explanatory note at the bottom of the poem? and the title at the beginning

Kiss Poem with Fallacious Argument

Document · Version 18

I've created a poem that demonstrates cacosistaton by starting with a potentially positive comparison and then systematically making it progressively worse. The kiss begins as "like a summer's burning day" but through repeated qualifications becomes destructive fire, then scorching desert heat, and finally furnace-breath. Each revision escalates the negative intensity, showing how this rhetorical figure can transform what might initially seem like praise into increasingly harsh criticism through a pattern of worsening qualifications.

add to the wisdom of literary expression by writing a poem with 6 to 8 lines about a kiss. please be sure to use an example of the word schema or figure known as PSEUDOMENOS in the poem and add the name of the figure or schema or trope (paranthetically) into the title like, "kiss bliss (with "the name of the figure')" also can you put and explanatory note at the bottom of the poem? and the title at the beginning

Kiss Poem with Fallacious Argument

Document · Version 19

I've created a poem that demonstrates pseudomenos (the liar's paradox) through a beloved who claims to always lie, creating a self-referential logical contradiction. If they always lie, then their statement about always lying must be false, which means they sometimes tell the truth. The poem explores this paradox in a romantic context, suggesting that the logical impossibility of the claim might actually create space for genuine expressions of love, even from someone who insists they always deceive.

add to the wisdom of literary expression by writing a poem with 6 to 8 lines about a kiss. please be sure to use an example of the word schema or figure known as ASSISTATON in the poem and add the name of the figure or schema or trope (paranthetically) into the title like, "kiss bliss (with "the name of the figure')" also can you put and explanatory note at the bottom of the poem? and the title at the beginning

Kiss Poem with Fallacious Argument

Document · Version 20

I've created a poem that demonstrates assistaton through the repetitive accumulation of comparisons all emphasizing the single quality of "sweetness." The poem builds layer upon layer of sweet comparisons—honey, fruit, songs, wine, lullaby, dawn, forgiveness—creating an overwhelming, almost hypnotic effect. This rhetorical figure doesn't just describe the kiss as sweet, but saturates the reader with multiple dimensions of sweetness through repetitive accumulation, making the quality seem all-encompassing and intensifying the emotional impact through sheer abundance of similar expressions.

add to the wisdom of literary expression by writing a poem with 6 to 8 lines about a kiss. please be sure to use an example of the word schema or figure known as CERATIN in the poem and add the name of the figure or schema or trope (paranthetically) into the title like, "kiss bliss (with "the name of the figure')" also can you put and explanatory note at the bottom of the poem? and the title at the beginning

I need to clarify something about "ceratin" - this doesn't appear to be a recognized rhetorical figure or logical fallacy in classical rhetoric or modern logic. I've searched through various sources on rhetorical figures, logical fallacies, and literary devices, but "ceratin" doesn't appear in any standard taxonomies. Could you clarify what you mean by "ceratin"? Are you perhaps thinking of:

- A specific rhetorical figure from a particular tradition?
- A variation or alternative name for another figure?
- Something from a specialized or regional system of rhetoric?

If you could provide more context about what "ceratin" involves or where you encountered this term, I'd be happy to create a poem that demonstrates the specific rhetorical or logical pattern you have in mind, with the proper title and explanatory note.

add to the wisdom of literary expression by writing a poem with 6 to 8 lines about a kiss. please be sure to

use an example of the word schema or figure known as THE HORNED ARGUMENT in the poem and add the name of the figure or schema or trope (paranthetically) into the title like, "kiss bliss (with "the name of the figure')" also can you put and explanatory note at the bottom of the poem? and the title at the beginning

Kiss Poem with Fallacious Argument

Document · Version 21

I've created a poem that demonstrates the horned argument (dilemma) by presenting two possible explanations for why a former lover still kisses skillfully, both of which lead to painful conclusions. The speaker is caught on the "horns" of this dilemma: either the beloved has been practicing with others (betrayal) or remembers their past intimacy too well (heartbreak). This rhetorical figure creates a logical trap where both alternatives result in emotional damage, forcing a confrontation with the impossibility of finding a satisfactory explanation.

add to the wisdom of literary expression by writing a poem with 6 to 8 lines about a kiss. please be sure to use an example of the word schema or figure known as THE CROCODILE'S ARGUMENT in the poem and add the name of the figure or schema or trope (paranthetically) into the title like, "kiss bliss (with "the name of the figure')" also can you put and explanatory note at the bottom of the poem? and the title at the beginning

Kiss Poem with Fallacious Argument

Document · Version 22

I've created a poem that demonstrates the crocodile's argument, a classical paradox where a promise is made contingent upon correctly guessing the promiser's true intention. The beloved promises to kiss only if the speaker can correctly guess whether they actually want to kiss, creating a logical trap. If the speaker guesses "yes," the kiss becomes obligatory rather than desired; if they guess "no," the beloved faces a contradiction in how to respond. This paradox shows how certain conditional promises create self-defeating logical structures that make genuine resolution impossible.

add to the wisdom of literary expression by writing a poem with 6 to 8 lines about a kiss. please be sure to use an example of the word schema or figure known as DISPUTATION in the poem and add the name of the figure or schema or trope (paranthetically) into the title like, "kiss bliss (with "the name of the figure')" also can you put and explanatory note at the bottom of the poem? and the title at the beginning

Kiss Poem with Fallacious Argument

Document · Version 23

I've created a poem that demonstrates disputation through a formal debate structure between two lovers about the meaning of their kiss. One argues from a materialist/scientific perspective (chemistry, neurons, dopamine) while the other takes a romantic/spiritual stance (souls connecting, true love, poetic authority). This creates a structured argument with opposing viewpoints, each citing their preferred evidence. The poem resolves by suggesting that their physical connection settles the intellectual dispute in a way that formal debate cannot, showing how disputation sometimes yields to more direct forms of understanding.

add to the wisdom of literary expression by writing a poem with 6 to 8 lines about a kiss. please be sure to use an example of the word schema or figure known as THE PROVE A PARADOX in the poem and add the name of the figure or schema or trope (paranthetically) into the title like, "kiss bliss (with "the name of the figure')" also can you put and explanatory note at the bottom of the poem? and the title at the beginning

Kiss Poem with Fallacious Argument

Document · Version 24

I've created a poem that demonstrates "the prove a paradox" by constructing a seemingly logical argument that leads to an absurd conclusion contradicting common sense. The speaker reasons that kissing someone proves you don't love them, because true love would require avoiding any risk to their health. This paradoxical proof takes a reasonable premise (caring about someone's wellbeing) to an extreme logical conclusion that defies intuitive understanding, showing how rigid reasoning can sometimes arrive at results that contradict both emotional truth and lived experience.

add to the wisdom of literary expression by writing a poem with 6 to 8 lines about a kiss. please be sure to use an example of the word schema or figure known

as APORIA in the poem and add the name of the figure or schema or trope (paranthetically) into the title like, "kiss bliss (with "the name of the figure')" also can you put and explanatory note at the bottom of the poem? and the title at the beginning

Kiss Poem with Fallacious Argument

Document · Version 25

I've created a poem that demonstrates aporia through genuine expressions of doubt and perplexity about how to adequately describe a kiss. The speaker is caught in authentic uncertainty, asking a series of questions that reveal their struggle with the inadequacy of language: "How can I speak...?" "What words could capture...?" "Should I compare...?" This creates the characteristic effect of aporia—being genuinely at a loss between the impossibility of adequate expression and the compulsion to attempt it anyway, showing the speaker's real bewilderment rather than rhetorical posturing.

add to the wisdom of literary expression by writing a poem with 6 to 8 lines about a kiss. please be sure to use an example of the word schema or figure known as ANTHYPOPHORA in the poem and add the name of the figure or schema or trope (paranthetically) into the title like, "kiss bliss (with "the name of the figure')" also can you put and explanatory note at the bottom of the poem? and the title at the beginning

Kiss Poem with Fallacious Argument

Document · Version 26

I've created a poem that demonstrates anthypophora through a structured pattern of the speaker asking questions and then immediately providing their own answers. The poem moves through a series of "What makes...?" and "What is it...?" questions, each followed by the speaker's response, building toward a final revelation about what makes this particular kiss special. This rhetorical figure allows the speaker to guide the reader through their thought process in an organized way, using the question-answer format as a deliberate structural device rather than expressing genuine uncertainty.

add to the wisdom of literary expression by writing a poem with 6 to 8 lines about a kiss. please be sure to use an example of the word schema or figure known as ANACOENOSIS in the poem and add the name of the figure or schema or trope (paranthetically) into the title like, "kiss bliss (with "the name of the figure')" also can you put and explanatory note at the bottom of the poem? and the title at the beginning

Kiss Poem with Fallacious Argument

Document · Version 27

I've created a poem that demonstrates anacoenosis by consistently appealing to the beloved's judgment and wisdom throughout. The speaker uses phrases like "What think you," "Tell me, dear friend, am I not right," and "Judge for yourself" to invite collaborative deliberation rather than imposing their own viewpoint. This rhetorical figure creates intimacy by treating the beloved as a trusted advisor whose opinion matters, building a sense of shared decision-making about whether to kiss under the starlit sky rather than simply trying to persuade through one-sided argument.

add to the wisdom of literary expression by writing a poem with 6 to 8 lines about a kiss. please be sure to use an example of the word schema or figure known as synchoresis in the poem and add the name of the figure or schema or trope (paranthetically) into the title like, "kiss bliss (with "the name of the figure')" also can you put and explanatory note at the bottom of the poem? and the title at the beginning

Kiss Poem with Fallacious Argument

Document · Version 28

I've created a poem that demonstrates synchoresis through strategic concessions that ultimately strengthen the speaker's position. The speaker grants multiple points to their opponent—acknowledging that their kiss was "rash," "unwise," "foolish," and "too fast"—using conciliatory language like "I grant you this," "Yes, we were," and "You're right to say." However, after making these strategic admissions, the speaker pivots to argue that despite these flaws, the experience was still worthwhile. This rhetorical technique builds credibility through apparent reasonableness while actually setting up a stronger counterargument.

add to the wisdom of literary expression by writing a poem with 6 to 8 lines about a kiss. please be sure to use an example of the word schema or figure known as PROCATALEPSIS in the poem and add the name of the figure or schema or trope (paranthetically) into the title like, "kiss bliss (with "the name of the figure')" also can you put and explanatory note at the bottom of the poem? and the title at the beginning

Kiss Poem with Fallacious Argument

Document · Version 29

I've created a poem that demonstrates procatalepsis by anticipating and preemptively addressing potential objections to the significance of a kiss. The speaker uses "Before you say" and "Before you claim" to acknowledge likely counterarguments—that the kiss was meaningless, wine-induced, or born of loneliness—then systematically responds to these anticipated criticisms. This rhetorical strategy strengthens the speaker's position by showing they've considered opposing viewpoints and can address them proactively, rather than being caught off-guard or appearing naive about potential objections.

add to the wisdom of literary expression by writing a poem with 6 to 8 lines about a kiss. please be sure to use an example of the word schema or figure known as PAROMOLOGIA in the poem and add the name of the figure or schema or trope (paranthetically) into the title like, "kiss bliss (with "the name of the figure')" also can you put and explanatory note at the bottom of the poem? and the title at the beginning

Kiss Poem with Fallacious Argument

Document · Version 30

I've created a poem that demonstrates paromologia by having the speaker openly admit to various accusations about their scandalous kiss—that it broke rules, was dangerous, foolish, and publicly inappropriate. Rather than defending against these charges, the speaker embraces them with frank acknowledgment ("I'll not deny," "Indeed," "yes, perhaps"). However, the rhetorical power comes in the final turn where these admitted faults are transformed into evidence of authentic passion, with the speaker declaring they'd gladly "sin again," turning the concessions into a strength rather than weakness.

add to the wisdom of literary expression by writing a poem with 6 to 8 lines about a kiss. please be sure to use an example of the word schema or figure known as concessio in the poem and add the name of the figure or schema or trope (paranthetically) into the title like, "kiss bliss (with "the name of the figure')" also can you put and explanatory note at the bottom of the poem? and the title at the beginning

Kiss Poem with Fallacious Argument

Document · Version 31

I've created a poem that demonstrates concessio through formal concessive language where the speaker systematically grants points to their opponent using phrases like "I grant," "Suppose I yield," and "Admit I broke." The speaker acknowledges that their kiss was wrong, improper, poorly timed, and damaging to their reputation. However, the rhetorical power emerges when the speaker pivots with "Still, even granting" to argue that despite accepting all these criticisms, the experience was still worthwhile. This shows how concessio can actually strengthen an argument by demonstrating that even the opponent's strongest points don't undermine the speaker's core position.

add to the wisdom of literary expression by writing a poem with 6 to 8 lines about a kiss. please be sure to use an example of the word schema or figure known as METASTASIS in the poem and add the name of the figure or schema or trope (paranthetically) into the title like, "kiss bliss (with "the name of the figure')" also can you put and explanatory note at the bottom of the poem? and the title at the beginning

Kiss Poem with Fallacious Argument

Document · Version 32

I've created a poem that demonstrates metastasis by systematically deflecting accusations back onto the accuser. Instead of defending against charges of improper haste and poor taste, the speaker redirects each criticism with counter-questions that implicate the beloved's own behavior: "who first lingered," "who sent letters," "what made you stay," "why did you not flee?" This rhetorical strategy shifts responsibility away from the speaker by suggesting that the accuser's own actions enabled or invited the very behavior they're now criticizing, effectively turning the tables on the source of the accusations.

add to the wisdom of literary expression by writing a poem with 6 to 8 lines about a kiss. please be sure to use an example of the word schema or figure known as apodioxis in the poem and add the name of the figure or schema or trope (paranthetically) into the title like, "kiss bliss (with "the name of the figure')" also can you put and explanatory note at the bottom of the poem? and the title at the beginning

Kiss Poem with Fallacious Argument

Document · Version 33

I've created a poem that demonstrates apodioxis through vehement rejection and contemptuous dismissal of opposing arguments. The speaker uses phrases like "Away with talk," "Let others prattle," "I'll hear no more," and "Enough of warnings" to dismiss critics without engaging their actual arguments. Rather than addressing concerns about propriety or hasty love, the speaker contemptuously labels critics as "gossips," "fools," and possessors of "cold and loveless force." This rhetorical figure creates a tone of superior disdain that refuses to dignify opposing viewpoints with serious consideration, instead sweeping them aside as unworthy of response.

add to the wisdom of literary expression by writing a poem with 6 to 8 lines about a kiss. please be sure to use an example of the word schema or figure known as diasyrmus in the poem and add the name of the figure or schema or trope (paranthetically) into the title like, "kiss bliss (with "the name of the figure')" also can you put and explanatory note at the bottom of the poem? and the title at the beginning

Kiss Poem with Fallacious Argument

Document · Version 34

I've created a poem that demonstrates diasyrmus through bitter irony and sarcastic praise that means the opposite of what's literally stated. The speaker uses harsh negative terms like "terrible," "dreadful," "unbearable," "awful," "horribly," and "ghastly" to describe the kiss, but the context makes it clear these are ironic reversals—the kiss was actually wonderful. This rhetorical figure works through the stark contrast between the condemning language chosen and the obviously positive effects described, creating a mocking tone that rejects criticism by sarcastically adopting the vocabulary of disparagement while making the true meaning unmistakable.

add to the wisdom of literary expression by writing a poem with 6 to 8 lines about a kiss. please be sure to use an example of the word schema or figure known as antirrhesis in the poem and add the name of the figure or schema or trope (paranthetically) into the title like, "kiss bliss (with "the name of the figure')" also can you put and explanatory note at the bottom of the poem? and the title at the beginning

Kiss Poem with Fallacious Argument

Document · Version 35

I've created a poem that demonstrates antirrhesis through direct contradiction and refutation of the opponent's statements. The beloved offers materialistic explanations for their kiss (desire, chemistry, hormones, evolution), which the speaker firmly rejects with explicit contradictions like "But I reject this claim entire" and "No—what we shared transcends." This rhetorical figure follows the pattern of stating the opponent's position, then strongly contradicting it, before presenting an alternative viewpoint. The speaker counters each scientific explanation with arguments for spiritual and transcendent interpretations of their intimate connection.

add to the wisdom of literary expression by writing a poem with 6 to 8 lines about a kiss. please be sure to use an example of the word schema or figure known as APHORISMUS in the poem and add the name of the figure or schema or trope (paranthetically) into the title like, "kiss bliss (with "the name of the figure')" also can you put and explanatory note at the bottom of the poem? and the title at the beginning

Kiss Poem with Fallacious Argument

Document · Version 36

I've created a poem that demonstrates aphorismus by rejecting the conventional meaning of "kiss" and proposing a more personal, profound definition. The speaker begins by dismissing the common saying "a kiss is just a kiss," then critiques how words like "love" and "passion" have been degraded from their true meanings. The speaker ultimately redefines what "kiss" means in their own context—not the shallow common usage, but something "deeper, rarer, more sincere." This rhetorical figure allows the reclaiming of language from its degraded popular usage and the restoration of more authentic, meaningful definitions.

add to the wisdom of literary expression by writing a poem with 6 to 8 lines about a kiss. please be sure to use an example of the word schema or figure known as COMMORATIO in the poem and add the name of the figure or schema or trope (paranthetically) into the title like, "kiss bliss (with "the name of the figure')" also can you put and explanatory note at the bottom of the poem? and the title at the beginning

Kiss Poem with Fallacious Argument

Document · Version 37

I've created a poem that demonstrates commoratio by dwelling extensively on a single kiss through repetitive elaboration. The speaker returns obsessively to "that kiss" with varied restatements—"the kiss that stopped all time," "that single kiss," "that perfect kiss," "one kiss, just one," "the kiss that changed," "that tender kiss." Each repetition adds a new dimension to the same central experience, creating an accumulative emphasis that suggests the speaker cannot move beyond this pivotal moment. This rhetorical figure shows how persistent elaboration of a single point can create intense focus and reveal the seemingly endless significance found in one transformative experience.

add to the wisdom of literary expression by writing a poem with 6 to 8 lines about a kiss. please be sure to use an example of the word schema or figure known as EPIMONE in the poem and add the name of the figure or schema or trope (paranthetically) into the title like, "kiss bliss (with "the name of the figure')" also can you put and explanatory note at the bottom of the poem? and the title at the beginning

Kiss Poem with Fallacious Argument

Document · Version 38

I've created a poem that demonstrates epimone through the relentless repetition of "Kiss me again" at the beginning of every line. This creates an urgent, almost desperate rhythm that mirrors the speaker's overwhelming desire. Unlike simple anaphora, epimone carries emotional urgency and persistence, suggesting the speaker cannot move beyond this single, consuming request. The repetition creates a hypnotic, pleading quality that emphasizes the intensity of need and the inability to be satisfied or distracted from this one overwhelming desire, making the repeated phrase both a plea and a demand.

add to the wisdom of literary expression by writing a poem with 6 to 8 lines about a kiss. please be sure to use an example of the word schema or figure known as APOPLANESIS in the poem and add the name of the figure or schema or trope (paranthetically) into the title like, "kiss bliss (with "the name of the figure')" also can you put and explanatory note at the bottom of the poem? and the title at the beginning

Kiss Poem with Fallacious Argument

Document · Version 39

Kiss Poem with Fallacious Argument

Document · Version 40

I've created a poem that demonstrates apoplanesis by having the speaker repeatedly deflect direct questions about their kiss by responding with seemingly irrelevant observations about nature. Instead of answering questions about motives, consequences, and meaning, the speaker diverts attention to blooming roses, shining stars, and moving leaves. These deflections appear random but serve to avoid concrete answers while suggesting that some experiences transcend rational explanation. This rhetorical figure creates a dreamy, evasive quality that refuses direct engagement with logical inquiry, implying that true intimacy exists beyond the reach of analytical questioning.

add to the wisdom of literary expression by writing a poem with 6 to 8 lines about a kiss. please be sure to use an example of the word schema or figure known as DICAEOLOGIA in the poem and add the name of the figure or schema or trope (paranthetically) into the title like, "kiss bliss (with "the name of the figure')" also can you put and explanatory note at the bottom of the poem? and the title at the beginning

Kiss Poem with Fallacious Argument

Document · Version 41

I've created a poem that demonstrates dicaeologia by having the speaker provide virtuous justifications for their intimate actions. Rather than denying the kiss, the speaker systematically offers noble explanations: kissing "only to console," holding close "just to make you whole," acting through "mercy" and "kindness" rather than passion. This rhetorical figure creates a pattern of self-justification that attempts to reframe potentially improper behavior as charitable actions, claiming pure motives for what are clearly intimate moments. The effect suggests someone perhaps protesting too much about the innocent nature of their obviously romantic behavior.

add to the wisdom of literary expression by writing a poem with 6 to 8 lines about a kiss. please be sure to use an example of the word schema or figure known as ANANGEON in the poem and add the name of the figure or schema or trope (paranthetically) into the title like, "kiss bliss (with "the name of the figure')" also can you put and explanatory note at the bottom of the poem? and the title at the beginning

Kiss Poem with Fallacious Argument

Document · Version 42

I've created a poem that demonstrates anangeon by having the speaker claim their kiss was compelled by forces beyond their control. The speaker systematically denies personal agency with phrases like "I had no choice," "The moonlight forced," "fate hung in the air," "My will was not my own," and "Necessity compelled." This rhetorical figure attributes intimate actions to external forces—destiny, celestial influence, unavoidable circumstances—rather than personal desire or decision. The effect allows the speaker to acknowledge their romantic behavior while deflecting responsibility by claiming they were merely instruments of larger, irresistible forces, removing personal culpability from what was clearly a chosen act.

add to the wisdom of literary expression by writing a poem with 6 to 8 lines about a kiss. please be sure to use an example of the word schema or figure known

as PAREURESIS in the poem and add the name of the figure or schema or trope (paranthetically) into the title like, "kiss bliss (with "the name of the figure')" also can you put and explanatory note at the bottom of the poem? and the title at the beginning

Kiss Poem with Fallacious Argument

Document · Version 43

I've created a poem that demonstrates pareuresis by having the speaker offer seemingly reasonable pretexts for romantic actions they clearly want to perform for entirely different reasons. The speaker provides rational-sounding explanations: kissing because lips were "parted," holding close to "deduce" feelings, whispering love to "explain" reactions, and touching to "test" emotions. These pretexts create a veneer of logical justification for what are obviously acts of desire and affection. This rhetorical figure allows engagement in romantic behavior while maintaining the fiction that each action serves some practical purpose rather than arising from simple emotional longing.

add to the wisdom of literary expression by writing a poem with 6 to 8 lines about a kiss. please be sure to use an example of the word schema or figure known as PYSMA in the poem and add the name of the figure or schema or trope (paranthetically) into the title like, "kiss bliss (with "the name of the figure')" also can you put and explanatory note at the bottom of the poem? and the title at the beginning

Kiss Poem with Fallacious Argument

Document · Version 44

I've created a poem that demonstrates pysma through earnest inquiry directed toward natural elements, seeking guidance about the moral and emotional complexities of a kiss. The speaker genuinely questions the moon, stars, wind, and earth about whether their passion was right or wrong, representing authentic uncertainty rather than rhetorical posturing. Unlike simple rhetorical questions, pysma shows real seeking for wisdom from higher or external sources. The figure creates a tone of sincere spiritual searching, though the silence that meets these inquiries suggests that some fundamental questions about love and desire must ultimately be answered from within rather than through external guidance.

add to the wisdom of literary expression by writing a poem with 6 to 8 lines about a kiss. please be sure to use an example of the word schema or figure known as THE FALLACY OF MANY QUESTIONS in the poem and add the name of the figure or schema or trope (paranthetically) into the title like, "kiss bliss (with "the name of the figure')" also can you put and explanatory note at the bottom of the poem? and the title at the beginning

Kiss Poem with Fallacious Argument

Document · Version 45

I've created a poem that demonstrates the fallacy of many questions through loaded questions that contain unproven assumptions. The questions "When will you stop pretending that you care?" and "why do you keep lying with each kiss?" both presuppose that the speaker is pretending and lying, forcing them to accept these false premises in order to answer directly. The poem shows the speaker recognizing this logical trap and refusing to engage with questions that embed unestablished accusations, instead challenging the false assumptions built into the interrogation itself. This illustrates how loaded questions can be used manipulatively to corner someone into accepting guilt or fault that hasn't been proven.

add to the wisdom of literary expression by writing a poem with 6 to 8 lines about a kiss. please be sure to use an example of the word schema or figure known as DISSUASION in the poem and add the name of the figure or schema or trope (paranthetically) into the title like, "kiss bliss (with "the name of the figure')" also can you put and explanatory note at the bottom of the poem? and the title at the beginning

Kiss Poem with Fallacious Argument

Document · Version 46

I've created a poem that demonstrates dissuasion by systematically arguing against kissing through multiple persuasive strategies. The speaker presents various discouraging reasons: fear of social scandal, moral concerns about reputation, historical examples of broken hearts, and warnings about future consequences. This creates a sustained rhetorical effort to prevent romantic action by cataloging potential negative outcomes. The figure works through accumulated warnings and appeals to prudence, though the very urgency and specificity of these deterrents might paradoxically reveal the speaker's own conflicted desires about what they're ostensibly trying to prevent.

add to the wisdom of literary expression by writing a poem with 6 to 8 lines about a kiss. please be sure to use an example of the word schema or figure known as PERSUASION in the poem and add the name of the figure or schema or trope (paranthetically) into the title like, "kiss bliss (with "the name of the figure')" also can you put and explanatory note at the bottom of the poem? and the title at the beginning

Kiss Poem with Fallacious Argument

Document · Version 47

I've created a poem that demonstrates persuasion through systematic encouragement of kissing using multiple rhetorical strategies. The speaker employs aesthetic appeal ("moonlight gilds the trees"), temporal urgency ("Life's fleeting moments," "each heartbeat quickly flees"), emotional reward ("joy that one small kiss could bring"), fear of loss ("sweetness we might never know again"), and metaphorical warning ("Love's opportunities take sudden wing"). This creates a sustained argument for immediate romantic action, building momentum through accumulated appeals to beauty, mortality, pleasure, and scarcity designed to overcome hesitation and move the beloved toward the desired decision to kiss.

add to the wisdom of literary expression by writing a poem with 6 to 8 lines about a kiss. please be sure to use an example of the word schema or figure known

as DEHORTATION in the poem and add the name of the figure or schema or trope (paranthetically) into the title like, "kiss bliss (with "the name of the figure')" also can you put and explanatory note at the bottom of the poem? and the title at the beginning

Kiss Poem with Fallacious Argument

Document · Version 48

I've created a poem that demonstrates dehortation through earnest pleading against kissing, despite the speaker's obvious desire for it. The speaker uses tender address ("My dearest love") while desperately counseling against action ("please do not," "Refrain," "I beg you," "Stay distant"), acknowledging their own passion while warning of serious consequences ("would undo us both," "bring destruction"). This rhetorical figure differs from simple dissuasion by carrying genuine concern and emotional investment—the speaker isn't arguing against something they don't want, but pleading against something they desperately desire but believe would be harmful, creating the characteristic tension of conflicted, heartfelt counsel.

add to the wisdom of literary expression by writing a poem with 6 to 8 lines about a kiss. please be sure to use an example of the word schema or figure known as PROTROPE in the poem and add the name of the figure or schema or trope (paranthetically) into the title like, "kiss bliss (with "the name of the figure')" also can you put and explanatory note at the bottom of the poem? and the title at the beginning

Kiss Poem with Fallacious Argument

Document · Version 49

Kiss Poem with Fallacious Argument

Document · Version 50

I've created a poem that demonstrates protrope through energetic encouragement and exhortation to pursue kissing. The speaker uses dynamic commands ("Come," "Embrace," "Rise up," "Take courage," "Cast off," "Move forward") combined with inspiring imagery and motivational appeals. Unlike simple persuasion, protrope has a rallying quality that seeks to rouse the listener from hesitation into bold action. The figure creates a tone of supportive urgency—like a coach or encouraging friend—using imperative verbs and uplifting language to inspire the beloved to overcome reservations and act on their desires, emphasizing empowerment and confident forward movement.

Kiss Bliss (with False Dilemma)

Your lips must choose: surrender all or flee—

No middle ground exists between our souls.

Either this kiss will bind us endlessly,

Or we remain forever incomplete wholes.

But in that instant when our mouths collide,

All logic fails, and hearts need not decide.

The sweetest truth lies not in either-or,

But in the tender space between what was and what's in store.

Explanatory Note: This poem employs the fallacious argument known as "false dilemma" (also called false dichotomy or either-or fallacy) in lines 1-4, where the speaker presents only two extreme options—total surrender or complete retreat—ignoring the nuanced possibilities that exist between these extremes. The poem then resolves this logical fallacy in the final lines by acknowledging that love and intimacy exist in the subtle spaces between absolutes, suggesting that the most profound human experiences often transcend binary thinking.

Kiss Bliss (with Equivocation)

Your kiss is like a bank—it holds my heart,

And I withdraw such interest from your lips.

You've made deposits of the sweetest art,

While passion compounds with each tender eclipse.

But banks are cold, and you are burning fire,

Your mouth a vault that feeds my soul's desire.

This tender commerce transcends all accounting—

Love's currency beyond any mortal counting.

Explanatory Note: This poem employs the material fallacy of "equivocation," where the word "bank" shifts meaning throughout the extended metaphor. Initially used in its financial sense (holding deposits, withdrawals, interest), it then conflicts with the emotional reality being described, as the speaker acknowledges that banks are "cold" while the beloved is "burning fire." This equivocation—using the same word with different meanings—creates both the poem's central conceit and its logical inconsistency, ultimately suggesting that love's reality exceeds the limitations of any single metaphor.

Kiss Bliss (with Antanaclasis)

Your lips speak volumes without saying a word,

And when they speak against mine, soft and low,

They speak a language only hearts have heard—

This tender speech that makes my spirit glow.

In silence, how your mouth can speak so true,

Each kiss a conversation, me and you.

What words could never speak, your touch makes clear:

Love speaks most loudly when no words appear.

Explanatory Note: This poem employs "antanaclasis," a rhetorical figure that involves the repetition of a word with different meanings or in different senses. The word "speak" appears throughout the poem with shifting meanings: first as verbal communication ("speak volumes," "without saying a word"), then as physical contact ("speak against mine"), then as expressing emotion ("speak a language only hearts have heard"), and finally as metaphorical communication ("Love speaks most loudly"). This repetition with varied meanings creates layers of significance while exploring how a kiss communicates beyond the boundaries of ordinary language.

Kiss Bliss (with Fallacy of Amphibology)

I kissed you with my eyes closed yesterday,

And felt the world dissolve in tender fire.

Your lips met mine in that sweet, breathless way

That left us both consumed with pure desire.

But did I close my eyes to kiss you true,

Or kiss you truly with my eyes closed tight?

The grammar blurs like morning's golden dew—

What matters is we found each other's light.

Explanatory Note: This poem employs the "fallacy of amphibology," which arises from ambiguous grammatical construction that creates multiple possible interpretations. The line "I kissed you with my eyes closed" demonstrates this ambiguity—it could mean either "I closed my eyes while kissing you" or "I kissed you by using my closed eyes" (an impossible but poetically suggested action). The poem then explicitly acknowledges this grammatical ambiguity in lines 5-6, questioning whether the eyes were closed for the purpose of kissing or whether the kissing was somehow done with closed eyes, before resolving that such linguistic precision matters less than the emotional truth of the moment.

Kiss Bliss (with Fallacy of Composition)

Each cell within your lips is soft and warm,

Each nerve beneath holds lightning for my touch.

Your breath, each molecule, creates a storm—

So surely you, entire, must be too much

For mortal hearts to bear without breaking.

But when we kiss, I find the whole transcends

Its parts: not overwhelming, but awaking

Something gentle where all passion blends.

Explanatory Note: This poem demonstrates the "fallacy of composition," which incorrectly assumes that what is true of the parts must also be true of the whole. The speaker examines individual components of the beloved's lips (cells, nerves, breath molecules) and concludes that if each part is intense or overwhelming, then the whole person must be "too much" to bear. However, the poem's resolution reveals this logical error—the actual experience of kissing shows that the whole (the complete person and kiss) creates something entirely different from what the individual parts might suggest, something "gentle" rather than overwhelming.

Kiss Bliss (with Fallacy of Division)

Our kiss was perfect, flawless as a dream,

So every breath within must be divine,

Each heartbeat during that celestial gleam

Must pulse with starlight, every touch sublime.

But pulling back, I find my lips still human,

My pulse still mortal, breath still filled with doubt.

The kiss was magic, but I'm still just woman—

Perfection's whole need not perfect parts throughout.

Explanatory Note: This poem demonstrates the "fallacy of division," which incorrectly assumes that what is true of the whole must also be true of each individual part. The speaker begins by reasoning that because their kiss was perfect, every component within that moment (each breath, heartbeat, touch) must also be perfect or divine. However, the poem's resolution reveals this logical error—while the kiss as a complete experience was transcendent, the individual elements (the speaker's lips, pulse, breath) remain wonderfully, imperfectly human. This shows how profound experiences can emerge from ordinary parts without requiring those parts to be extraordinary themselves.

Kiss Bliss (with Fallacy of Form of Speech)

"Your kiss," I said, "is like a summer's day"—

But metaphors deceive with borrowed light.

I called it "honey sweet" in passion's play,

Yet honey cannot capture love's true sight.

"You taste like starlight," "lips of morning dew"—

Each phrase a pretty lie that falls apart.

No figure holds the simple, stunning truth:

Your kiss is just your kiss upon my heart.

Explanatory Note: This poem demonstrates the "fallacy of form of speech" (also known as the fallacy of figurative language), which occurs when figurative expressions are treated as if they were literal truths or when metaphors are mistaken for accurate descriptions of reality. The speaker employs various poetic comparisons ("summer's day," "honey sweet," "starlight," "morning dew") but then recognizes that these figures of speech, while beautiful, fail to capture the actual experience. The poem concludes by acknowledging that no metaphor can substitute for the direct, unadorned reality of the kiss itself, revealing how figurative language, despite its beauty, can sometimes obscure rather than illuminate truth.

Kiss Bliss (with Fallacy of Accent)

You whispered, "I will NEVER leave you here"—

The emphasis on "never" made me soar.

But later, in the light, I heard it clear:

"I will never LEAVE you here" meant more.

Not "never leave," but "never leave you HERE"—

In this sad place, but somewhere else instead.

One kiss can shift what words were meant to bear:

The accent changes everything that's said.

Explanatory Note: This poem demonstrates the "fallacy of accent," which occurs when the meaning of a statement is altered by changing the emphasis or stress placed on different words within the same sentence. The phrase "I will never leave you here" takes on completely different meanings depending on which word receives emphasis: stressing "NEVER" suggests permanent devotion, while stressing "LEAVE" implies abandonment is not the issue, and stressing "HERE" suggests departure from this specific location rather than departure altogether. The poem shows how a single kiss and moment of intimacy can cause us to misinterpret the intended accent or emphasis, leading to false conclusions about what was actually promised or meant.

Kiss Bliss (with Fallacy of Accident)

They say that kissing spreads disease and pain,

That lips are vectors for infection's course.

So surely ours must follow nature's reign

And poison us with some corrupted force.

But here we stand, months later, whole and well—

Our kiss defied what general rules foretell.

Some truths bend when the heart's exception proves

That love creates its own protective moves.

Explanatory Note: This poem demonstrates the "fallacy of accident" (also called "sweeping generalization" or "dicto simpliciter"), which occurs when a general rule or principle is applied to a specific case that should be treated as an exception. The speaker begins with the general medical principle that kissing can transmit illness, then incorrectly assumes this rule must apply universally to their specific situation. The poem's resolution suggests that their particular kiss represents an exception to the general rule, though it playfully attributes this to love's power rather than acknowledging that the general rule was never meant to apply to every single instance of kissing.

Kiss Bliss (with Fallacy of False Assumption or Secundum Quid)

Since all first kisses are awkward and shy,

And ours was graceful, tender, and complete,

This cannot be our first—that's just a lie

You told to make this moment seem more sweet.

But wait—perhaps my premise led astray:

Not every first kiss stumbles in the dark.

Some souls meet perfectly on their first day,

And leave upon each other lasting mark

Explanatory Note: This poem demonstrates the fallacy of "secundum quid" (also known as the fallacy of false assumption or hasty generalization), which occurs when a conclusion is drawn from insufficient evidence or when a qualified statement is treated as an absolute rule. The speaker begins with the assumption that "all first kisses are awkward," then uses this questionable generalization to conclude that their graceful kiss cannot possibly be a first kiss. The poem's resolution reveals the logical error—the initial assumption was too broad and didn't account for exceptions. This fallacy often involves taking limited experience or partial truths and applying them universally without sufficient justification.

Kiss Bliss (with Fallacy of Ignorance of the Elench)

You asked me if our kiss meant I would stay,

And I spoke of the beauty of your eyes,

The way your hair catches the morning's ray,

How sweet your voice sounds when it softly sighs.

But you pressed on: "Does kissing mean we're bound?"

I praised your gentle hands, your graceful walk.

Your question hangs between us like a sound

That echoes while I dance around your talk

Explanatory Note: This poem demonstrates "ignorance of the elench" (also called "ignoratio elenchi" or "missing the point"), a fallacy that occurs when someone responds to an argument or question by addressing something related but fundamentally different from what was actually asked. The beloved asks a direct question about commitment ("Does our kiss mean I would stay?" and "Does kissing mean we're bound?"), but the speaker consistently responds by praising physical attributes and avoiding the actual question. This evasive pattern shows how we sometimes substitute easier or more comfortable topics for the difficult questions we're actually being asked, thereby missing the point entirely.

Kiss Bliss (with Fallacy of the Consequent)

If you truly loved me, you would kiss me so,

And here you are, kissing me just that way.

Therefore your love for me must be true—

I know This kiss proves all the words you cannot say.
But wait—perhaps you kiss like this with all,

Or maybe practice makes your technique shine.

A kiss alone cannot prove love's sweet call:

The same effect need not mean hearts align.

Explanatory Note: This poem demonstrates the "fallacy of the consequent" (also called "affirming the consequent"), a logical error with the form: "If P, then Q. Q is true. Therefore, P is true." The speaker reasons: "If you loved me, you would kiss me this way. You are kissing me this way. Therefore, you love me." This reasoning is fallacious because the same outcome (a passionate kiss) could result from multiple different causes—practice, habit, physical attraction, or simply good technique—not necessarily love. The poem's resolution acknowledges this logical error, recognizing that identical effects can spring from entirely different motivations.

Kiss Bliss (with Fallacy of False Cause)

The moment that we kissed, the rain began,

The thunder rolled across the evening sky.

Our passion stirred the storm—I understand

How love can make the very heavens cry.

But then I think: perhaps the clouds were there

Before our lips had ever met in bliss.

The weather follows patterns, unaware

Of mortal hearts or any stolen kiss.

Explanatory Note: This poem demonstrates the "fallacy of false cause" (also called "post hoc ergo propter hoc" or "false correlation"), which incorrectly assumes that because two events occurred in sequence, the first must have caused the second. The speaker observes that a storm began immediately after their kiss and concludes that their passion must have triggered the weather. This fallacious reasoning ignores the more likely explanation that the storm was already developing according to natural meteorological patterns, independent of any human romantic activity. The poem's resolution acknowledges this logical error by recognizing that correlation in timing doesn't establish causation.

Kiss Bliss (with Fallacy of Begging the Question)

I know our kiss was meant to be because

We kissed with such a perfect, fated feel.

And how do I know fate was there? It was—

The kiss itself makes destiny seem real.

This circular dance of heart and mind

Spins logic into lovers' sweet pretense:

We're meant to be because we're meant to find

Each other's lips—the proof is evidence.

Explanatory Note: This poem demonstrates the fallacy of "begging the question" (also called "circular reasoning" or "petitio principii"), which occurs when the conclusion of an argument is assumed in its premises. The speaker argues that their kiss was destined because it felt fated, then proves it was fated by pointing to the destined feeling of the kiss. This creates a circular loop where the conclusion (they're meant to be) is used to support the premise (the kiss felt fated), which in turn supports the same conclusion. The poem acknowledges this logical circularity while suggesting that in matters of love, such reasoning, though fallacious, creates its own kind of meaningful "pretense."

Kiss Bliss (with Fallacy of Many Questions)

"Why did you kiss me if you plan to leave?"

You ask, your eyes ablaze with hurt and doubt.

But wait—I never said I would deceive

Or spoke of leaving; what's this talk about?

Your question traps me in a false design:

It asks for reasons for a choice unmade.

I kissed you simply because you were mine

In that sweet moment—no escape plan laid.

Explanatory Note: This poem demonstrates the "fallacy of many questions" (also called "complex question" or "loaded question"), which occurs when a question contains an unproven assumption that the respondent is forced to accept in order to answer. The question "Why did you kiss me if you plan to leave?" assumes that the speaker does indeed plan to leave, even though this assumption hasn't been established. By attempting to answer this question directly, the speaker would implicitly accept the premise that they intend to leave. The poem shows how the speaker recognizes this logical trap and refuses to be cornered into accepting the false assumption embedded within the question.

Kiss Bliss (with Antistrephon)

You say I kiss too softly for your taste,

But I say you kiss far too hard for mine.

You claim my gentle touch is time gone waste,

While I find your fierce passion too divine.

"Too soft!" you cry. "Too hard!" I answer back.

"Too slow!" "Too fast!" our accusations fly.

Yet in this sweet debate of give and lack,

Our arguing lips meet—and both sides lie.

Explanatory Note: This poem demonstrates "antistrephon," a rhetorical figure involving the turning back of an opponent's argument or accusation upon themselves, often through direct contradiction or counter-accusation. The structure shows each lover accusing the other of kissing incorrectly—too soft, too hard, too slow, too fast—with each criticism immediately met by an opposite counter-criticism. This creates a pattern of mutual contradiction where each person deflects blame by turning the same type of accusation back on their partner. The poem resolves by suggesting that this argumentative exchange itself becomes a form of intimate connection, making both accusations ultimately meaningless.

Kiss Bliss (with Cacosistaton)

Your kiss is like a summer's burning day—

No, worse than fire, it scorches and destroys.

It's not just warm, but blazing hot, I'd say—

Not pleasant heat, but pain that fire employs.

This kiss brings not the gentle summer's grace,

But desert heat that bakes and cracks the earth.

Not sunshine's kiss upon a lover's face,

But furnace-breath that proves destruction's worth.

Explanatory Note: This poem demonstrates "cacosistaton," a rhetorical figure that involves the repeated qualification and worsening of a comparison or description, where each subsequent phrase makes the situation sound progressively more severe or negative. The speaker begins with a seemingly positive comparison ("like a summer's burning day") but then systematically makes it worse with each line—moving from pleasant warmth to destructive fire, from gentle heat to desert conditions, from sunshine's kiss to furnace-breath. This creates an escalating pattern where each revision intensifies the negative aspects, turning what might have been a compliment into an increasingly harsh critique.

Kiss Bliss (with Pseudomenos)

"I never tell the truth," you whispered low,

Your lips against my ear in lovers' play.

"So when I say I love you, you should know

That every word I speak to you's a lie."

But if you lie in all you ever say,

Then saying that you lie must be untrue—

Which means you sometimes speak truth anyway,

And maybe "I love you" is honest too.

Explanatory Note: This poem demonstrates "pseudomenos" (also known as the "liar's paradox"), a logical paradox that creates a self-referential contradiction. The beloved claims to "never tell the truth" and that "every word I speak to you's a lie." However, if this statement itself is true, then they do sometimes tell the truth, which contradicts their claim. If the statement is false, then they don't always lie, which again contradicts the assertion. The poem explores this paradox in the context of a romantic declaration, suggesting that the logical contradiction might actually allow room for the possibility that expressions of love could be genuine, even from someone who claims to always lie.

Kiss Bliss (with Assistaton)

Your kiss is sweet—yes, sweet as honey's gold,

As sweet as summer fruit upon the vine,

Sweet as the songs that ancient poets told,

Sweet as the first taste of communion wine.

Sweet as a mother's lullaby at night,

Sweet as the dawn that ends a lover's pain,

Sweet as forgiveness after bitter fight—

Your kiss is sweet, and sweet, and sweet again.

Explanatory Note: This poem demonstrates "assistaton," a rhetorical figure that involves the repetition and accumulation of similar expressions or comparisons to emphasize and intensify a single quality or idea. The word "sweet" and its variations appear throughout the poem in an escalating series of comparisons—honey, fruit, songs, wine, lullaby, dawn, forgiveness—each building upon the previous one to create a mounting effect. This rhetorical strategy doesn't just state that the kiss is sweet; it overwhelms the reader with multiple dimensions of sweetness, creating an almost hypnotic emphasis through repetitive accumulation that makes the quality seem all-encompassing and undeniable.

Kiss Bliss (with The Horned Argument)

You have not lost the art of kissing yet—

So either you still practice on another,

Or you remember what we can't forget:

The way we kissed when we were still together.

If you practice still, then you betray my trust,

If you remember well, you break my heart.

Either way, this kiss we share must

Leave one of us torn completely apart.

Explanatory Note: This poem demonstrates "the horned argument" (also called a "dilemma"), a rhetorical figure that presents two alternatives, both of which lead to an undesirable conclusion, like being caught on the horns of a bull. The speaker observes that their former lover still kisses skillfully, which leads to two possible explanations: either they've been practicing with someone else (betrayal) or they remember their past intimacy perfectly (heartbreak). Both "horns" of this dilemma lead to painful outcomes, creating a logical trap where any explanation for the beloved's kissing ability results in emotional damage. This figure forces the audience to confront the impossibility of a satisfactory resolution.

Kiss Bliss (with The Crocodile's Argument)

"I'll kiss you," you said, "if you guess right

Whether I truly want to kiss you now.

But if you guess wrong, I'll bid you goodnight

And never let these lips touch yours somehow."

I thought: if I say yes, you want to kiss,

But then you'd have to—so your want's not true.

If no, you don't want to, but you'd still miss

The chance to prove your paradox right through.

Explanatory Note: This poem demonstrates "the crocodile's argument" (or "crocodile dilemma"), a classical paradox in which a promise is made contingent upon correctly guessing the promiser's intention, creating a logical trap. The beloved promises to kiss if the speaker correctly guesses whether they want to kiss, but will refuse if the guess is wrong. This creates a paradox: if the speaker guesses "yes, you want to kiss me," then the beloved must follow through, making their desire artificial; if the speaker guesses "no, you don't want to," then the beloved could either prove them right (by not kissing) or wrong (by kissing anyway), but either response undermines the logical structure of the original proposition.

Kiss Bliss (with Disputation)

You argue that our kiss means nothing more

Than simple chemistry and fleeting lust.

I counter: souls connect through lips, what's more,

True love speaks louder than mere physical thrust.

You cite the science—neurons, dopamine,

While I invoke the poets and their art.

We debate the meaning while our lips convene

And settle arguments without our heart.

Explanatory Note: This poem demonstrates "disputation," a rhetorical figure involving formal debate or systematic argument between opposing viewpoints, often structured as a scholarly or logical exchange. The two lovers engage in a structured argument about the meaning of their kiss—one taking the materialist position (chemistry and lust) while the other argues for the spiritual interpretation (souls connecting through love). Each side presents evidence from their chosen authority (science versus poetry), creating a formal debate structure. The poem's resolution suggests that while their minds engage in intellectual disputation, their physical connection transcends and resolves the argument in a way that logical debate cannot.

Kiss Bliss (with The Prove a Paradox)

I'll prove that kissing you means I don't care:

For if I loved you truly, I would fear

To risk your health with germs that I might bear,

And keep my distance, holding you most dear.

But since I kiss you freely, without thought

Of danger that my lips might bring to you,

I clearly love you less than I once thought—

My careless kiss proves my affection's through.

Explanatory Note: This poem demonstrates "the prove a paradox" (also called "paradoxical proof"), a rhetorical figure that uses seemingly logical reasoning to arrive at a conclusion that contradicts common sense or expected reality. The speaker constructs an apparently rational argument that kissing someone proves you don't love them, reasoning that true love would require avoiding any risk to the beloved's health. This creates a paradox where the very act of intimate affection becomes evidence of its absence. The figure works by taking a premise to an extreme logical conclusion that defies intuitive understanding, showing how rigid logic can sometimes lead to absurd results that contradict lived experience and emotional truth.

Kiss Bliss (with Aporia)

How can I speak of what your kiss contains?

What words could capture lightning made of touch?

I know not how to tell you of these gains—

Is silence better? Have I said too much?

Should I compare it to the morning dew?

But what is dew beside your gentle fire?

I cannot say what kissing means to do—

Yet how can I not speak of such desire?

Explanatory Note: This poem demonstrates "aporia," a rhetorical figure expressing genuine doubt, perplexity, or uncertainty about how to proceed, often manifesting as a series of questions about the speaker's own ability to adequately express something. The speaker struggles with the fundamental impossibility of capturing the essence of a kiss in words, repeatedly questioning their approach: "How can I speak...?" "What words could capture...?" "Should I compare...?" "Yet how can I not speak...?" This creates a sense of being genuinely at a loss, caught between the inadequacy of language and the necessity of attempting expression. Aporia differs from rhetorical questions in that it represents authentic perplexity rather than disguised statements.

Kiss Bliss (with Anthypophora)

What makes your kiss so different from the rest?

It's not the softness of your willing lips.

What sets it apart from all that I've caressed?

It's not the way your gentle breathing slips.

What is it then that makes my heart catch fire?

Perhaps it's how you kiss me like you mean

To steal away my very soul's desire—

The most tender theft I've ever seen.

Explanatory Note: This poem demonstrates "anthypophora" (also called "hypophora"), a rhetorical figure in which the speaker poses a question and then immediately provides their own answer. The structure involves asking "What makes your kiss so different?" then answering "It's not the softness..." followed by another question "What sets it apart?" with its corresponding answer. This creates a pattern of self-interrogation and response that allows the speaker to guide the reader through their reasoning process, building toward the final revelation. Unlike aporia (genuine uncertainty) or simple rhetorical questions, anthypophora uses the question-answer format as a deliberate structural device to organize and present ideas.

Kiss Bliss (with Anacoenosis)

Should we not kiss beneath this starlit sky?

What think you of the wisdom of this choice?

Does not your heart, like mine, begin to fly

When moonlight gives our passion tender voice?

Tell me, dear friend, am I not right to feel

That kisses shared in darkness hold more fire?

Judge for yourself—does not this moment steal

Your breath away and kindle sweet desire?

Explanatory Note: This poem demonstrates "anacoenosis" (also called "communicatio"), a rhetorical figure in which the speaker asks for the audience's opinion or judgment, appealing to their wisdom and treating them as a collaborative partner in reaching a conclusion. The speaker consistently seeks the beloved's input with phrases like "What think you," "Tell me, dear friend, am I not right," and "Judge for yourself." This creates a sense of shared deliberation rather than one-sided persuasion, inviting the listener to participate actively in evaluating the situation. Anacoenosis builds intimacy and trust by treating the audience as capable judges whose opinion matters, rather than simply trying to impose a predetermined viewpoint.

Kiss Bliss (with Synchoresis)

I grant you this: our kiss was rash and wild,

Perhaps unwise for hearts so newly met.

Yes, we were foolish, passion-drunk and biled

With sweet abandon we might soon regret.

You're right to say we moved too fast, too free—

But oh, what rapture in that reckless flight!

I'll own our folly, yet still disagree:

Sometimes the sweetest wrongs feel most right.

Explanatory Note: This poem demonstrates "synchoresis," a rhetorical figure in which the speaker makes a strategic concession to their opponent's argument in order to strengthen their own position or build credibility before presenting their counterpoint. The speaker acknowledges multiple criticisms of their kiss—that it was "rash and wild," "unwise," "foolish," and "too fast"—using phrases like "I grant you this," "Yes, we were," and "You're right to say." However, after making these concessions, the speaker pivots with "But oh" and "yet still disagree" to argue that despite these admitted flaws, the experience was still worthwhile. This strategic acknowledgment of weaknesses actually strengthens the final position by showing reasonableness and honesty.

Kiss Bliss (with Procatalepsis)

Before you say our kiss was meaningless,

A fleeting moment sparked by wine and night,

Before you claim it was just loneliness

That drew our lips together in moonlight—

Consider this: though passion can deceive,

And yes, the wine may cloud our clearer sight,

Some truths transcend what cynics won't believe:

This kiss felt real, however wrong or right.

Explanatory Note: This poem demonstrates "procatalepsis" (also called "preemption" or "anticipation"), a rhetorical figure in which the speaker anticipates and addresses potential objections or counterarguments before they can be raised. The speaker begins with "Before you say" and "Before you claim," explicitly acknowledging likely criticisms—that the kiss was meaningless, driven by wine, or motivated by loneliness—and then systematically responds to these anticipated objections. This technique strengthens the speaker's position by showing awareness of opposing viewpoints and demonstrating that they have considered and can refute potential counterarguments, rather than ignoring or being blindsided by them.

Kiss Bliss (with Paromologia)

I'll not deny our kiss was scandalous—

Let gossips speak of how we broke the rules.

I won't pretend it wasn't dangerous,

Or claim we weren't a pair of lovesick fools.

Indeed, we kissed where all the world could see,

And yes, perhaps we should have shown restraint.

But having owned our sweet impropriety,

I'd sin again without the least complaint.

Explanatory Note: This poem demonstrates "paromologia," a rhetorical figure in which the speaker openly admits to accusations or criticisms, but then either diminishes their importance or turns them to advantage. The speaker freely acknowledges that their kiss was "scandalous," "dangerous," and that they were "lovesick fools" who kissed publicly without restraint. Rather than defending against these charges, the speaker embraces them with phrases like "I'll not deny," "I won't pretend," and "Indeed." However, after making these frank admissions, the speaker turns the concessions to their advantage by declaring they would "sin again without the least complaint," transforming the acknowledged faults into evidence of passionate authenticity.

Kiss Bliss (with Concessio)

I grant that kissing you was wholly wrong,

That propriety forbade our lips to meet.

Suppose I yield—our passion was too strong,

Our timing poor, our conduct indiscreet.

Admit I broke the rules of common sense,

And threw away my reputation's worth.

Still, even granting each grave consequence,

I'd trade it all again for moments' mirth.

Explanatory Note: This poem demonstrates "concessio," a rhetorical figure in which the speaker grants or yields points to their opponent, often using formal language of concession, but then proceeds to argue their position regardless of these admissions. The speaker uses explicit concessive language—"I grant," "Suppose I yield," "Admit I broke"—to acknowledge that their kiss was wrong, improper, poorly timed, and costly to their reputation. However, after making these formal concessions, the speaker pivots with "Still, even granting" to argue that despite all these acknowledged problems, the experience was worth the cost. This creates a powerful effect by showing that even accepting the opponent's strongest criticisms doesn't diminish the speaker's core position.

Kiss Bliss (with Metastasis)

You say I kissed you with improper haste—

But who first lingered when our eyes first met?

You claim my passion was in poor taste—

Yet who sent letters filled with sweet regret?

If I'm too bold, what made you stay so near?

If I'm too forward, why did you not flee?

The fault you find in me, my love, I fear

Lies deeper in your own heart's strategy.

Explanatory Note: This poem demonstrates "metastasis," a rhetorical figure in which the speaker deflects blame or responsibility by turning accusations back upon the accuser or by shifting the focus to external circumstances. Rather than defending against the beloved's accusations of improper haste and poor taste, the speaker systematically redirects each criticism back to the accuser's own behavior: "who first lingered," "who sent letters," "what made you stay," "why did you not flee?" This creates a pattern where every fault attributed to the speaker is reframed as originating from or being enabled by the accuser's own actions, effectively shifting responsibility away from the speaker and onto the very person making the accusations.

Kiss Bliss (with Apodioxis)

Away with talk of why we shouldn't kiss!

Let others prattle of propriety's rules—

I'll hear no more of what they call amiss,

Such chatter's fit for gossips and for fools.

Enough of warnings about love's swift course!

Your lips speak truths no sermon can refute.

Let critics keep their cold and loveless force—

Our kiss needs no defense, gives no dispute.

Explanatory Note: This poem demonstrates "apodioxis," a rhetorical figure involving the vehement rejection or dismissal of an opponent's argument, often with contemptuous language that refuses to engage with the substance of their position. The speaker uses dismissive phrases like "Away with talk," "Let others prattle," "I'll hear no more," and "Enough of warnings" to reject criticisms without addressing them directly. Rather than countering specific arguments about impropriety or hasty love, the speaker contemptuously dismisses the critics themselves as "gossips," "fools," and possessors of "cold and loveless force." This creates a tone of superior disdain that refuses to dignify opposing viewpoints with serious consideration.

Kiss Bliss (with Diasyrmus)

Oh yes, your kiss was truly terrible—

So soft it made my heart forget to beat.

How dreadful was that touch unbearable,

That left me dizzy, weak upon my feet.

What awful pleasure in your lips I found!

How horribly your breath made my pulse race!

Such ghastly bliss should surely be renowned

For ruining completely my embrace.

Explanatory Note: This poem demonstrates "diasyrmus," a rhetorical figure that involves rejecting or disparaging something through bitter irony or sarcastic praise that clearly means the opposite of what is literally stated. The speaker uses negative terms like "terrible," "dreadful," "unbearable," "awful," "horribly," and "ghastly" to describe the kiss, but the context makes clear these are ironic reversals—the kiss was actually wonderful. This creates a mocking tone that rejects criticism by ironically adopting the language of condemnation while making it obvious that the opposite meaning is intended. The figure works through the contrast between the harsh words chosen and the clearly positive effects described.

Kiss Bliss (with Antirrhesis)

"Your kiss," you say, "means nothing but desire—

Just chemistry and hormones in the blood."

But I reject this claim entire:

Love is not mere biological flood.

"It's simply evolution's ancient game,"

You argue with your cold, scientific mind.

No—what we shared transcends such sterile frame:

Our souls touched in a way none can define.

Explanatory Note: This poem demonstrates "antirrhesis," a rhetorical figure that involves the direct contradiction and refutation of an opponent's statement, often through explicit rejection followed by a counter-argument. The structure typically follows the pattern of stating the opponent's position, then firmly rejecting it with words like "No," "But I reject," or similar strong contradictions, before presenting an alternative viewpoint. Here, the beloved's materialistic explanations for their kiss (desire, chemistry, hormones, evolution) are each met with direct refutation ("But I reject this claim entire," "No—what we shared transcends") followed by the speaker's opposing interpretation that emphasizes spiritual and transcendent elements over purely physical explanations.

Kiss Bliss (with Aphorismus)

They say a kiss is just a kiss, but no—

A kiss can be a promise or goodbye.

When "love" means lust, then words lose all they owe

To truth; when "passion" fades to "getting by,"

The language dies. So when I kiss you here,

I mean not "kiss" as others use the word,

But something deeper, rarer, more sincere—

A vow that common speech has never heard.

Explanatory Note: This poem demonstrates "aphorismus," a rhetorical figure that involves rejecting the common or accepted meaning of a word and proposing a new, more precise, or personal definition in its place. The speaker begins by rejecting the conventional understanding that "a kiss is just a kiss," then goes on to distinguish between superficial meanings of words like "love" (which can mean mere lust) and "passion" (which can fade to routine) versus their deeper, more authentic meanings. The speaker ultimately redefines what "kiss" means in their personal context—not the common usage, but "something deeper, rarer, more sincere." This figure allows the speaker to reclaim language from its degraded common usage and restore more meaningful definitions.

Kiss Bliss (with Commoratio)

Your kiss—that kiss—the kiss that stopped all time,

That single kiss beneath the April sky,

The kiss that made my heartbeat lose its rhyme,

That perfect kiss that taught my soul to fly.

One kiss, just one, yet everything it held:

The kiss that changed the way I see the world,

That tender kiss where all my walls fell,

The kiss that left my very being unfurled.

Explanatory Note: This poem demonstrates "commoratio," a rhetorical figure that involves dwelling extensively on a single point or subject, often through repetitive elaboration and varied restatement of the same idea. The speaker focuses entirely on "that kiss," returning to it again and again with different descriptive phrases—"the kiss that stopped all time," "that single kiss," "that perfect kiss," "one kiss, just one," "the kiss that changed," "that tender kiss." Each repetition adds a new dimension or effect of the same central experience, creating an accumulative emphasis that suggests the speaker cannot move beyond this pivotal moment, finding endless significance in this single event through persistent, almost obsessive, elaboration.

Kiss Bliss (with Epimone)

Kiss me again—I need to feel your lips.

Kiss me again—the first was not enough.

Kiss me again—my heart still yearns and skips.

Kiss me again—one kiss feels far too rough.

Kiss me again—I'm drowning in this need.

Kiss me again—please don't deny my plea.

Kiss me again—your kiss is all I need.

Kiss me again—just one more, set me free.

Explanatory Note: This poem demonstrates "epimone," a rhetorical figure involving the frequent repetition of the same phrase or expression, often as an urgent plea or insistent demand. The phrase "Kiss me again" appears at the beginning of every line, creating a relentless, almost desperate rhythm that mirrors the speaker's overwhelming desire. Unlike simple anaphora (which repeats for structural effect), epimone carries an emotional urgency and persistence, suggesting the speaker cannot move beyond this single, consuming request. The repetition creates an almost hypnotic, pleading quality that emphasizes the intensity of the speaker's need and their inability to be satisfied or distracted from this one overwhelming desire.

Kiss Bliss (with Apoplanesis)

You ask me why I kissed you by the gate—

But look, the roses bloom despite the frost!

You want to know if this will be our fate—

See how the stars shine though the day is lost.

You press me for the meaning of that touch—

The wind still moves the leaves without a sound.

Some questions seek for answers far too much:

True kisses need no reasons to be found.

Explanatory Note: This poem demonstrates "apoplanesis," a rhetorical figure that involves deflecting or evading a direct question by responding with something seemingly irrelevant or by leading the conversation in an entirely different direction. Rather than answering the direct questions about motives ("why I kissed you"), consequences ("if this will be our fate"), and meaning ("the meaning of that touch"), the speaker repeatedly diverts attention to unrelated natural phenomena—blooming roses, shining stars, moving leaves. These deflections appear random but serve to avoid giving concrete answers while suggesting that some experiences, like kisses, transcend rational explanation. The figure creates a dreamy, evasive quality that refuses direct engagement with logical inquiry.

Kiss Bliss (with Dicaeologia)

I kissed you, yes, but only to console

Your tears that fell like April's gentle rain.

I held you close, but just to make you whole,

To ease your heart's unbearable sweet pain.

My lips found yours through mercy, nothing more—

A balm to heal the wounds that love had made.

I kissed to comfort, not to break love's law:

Pure kindness moved me, not some grand parade.

Explanatory Note: This poem demonstrates "dicaeologia," a rhetorical figure that involves making an excuse or providing justification for one's actions, often by claiming noble or innocent motives for behavior that might otherwise appear questionable. The speaker doesn't deny kissing their beloved, but systematically provides virtuous explanations for each intimate act: the kiss was "only to console," the embrace was "just to make you whole," the physical contact came from "mercy" and "kindness" rather than passion or desire. This creates a pattern of self-justification that attempts to reframe potentially improper behavior as charitable actions, suggesting pure motives while perhaps protesting too much about the innocent nature of clearly intimate moments.

Kiss Bliss (with Anangeon)

I had no choice but kiss you standing there—

The moonlight forced my lips to find your own.

What could I do when fate hung in the air?

My will was not my own, I was not shown

Another path. Necessity compelled

Each tender touch, each whispered word of love.

The stars themselves my actions have foretold:

I kissed because I had to, nothing of

Free choice remained when destiny took hold.

Explanatory Note: This poem demonstrates "anangeon" (also called "necessitas"), a rhetorical figure that involves claiming that one's actions were compelled by necessity, fate, or circumstances beyond personal control, thereby removing personal responsibility or blame. The speaker systematically denies agency in their romantic actions: "I had no choice," "The moonlight forced," "fate hung in the air," "My will was not my own," "Necessity compelled," "The stars themselves my actions have foretold." This creates a pattern where intimate behavior is attributed to external forces—destiny, celestial influence, or unavoidable circumstances—rather than personal desire or decision. The figure allows the speaker to acknowledge their actions while deflecting responsibility by claiming they were merely instruments of larger forces.

Kiss Bliss (with Pareuresis)

I kissed you—well, I found some small excuse:

Your lips were slightly parted as you spoke.

I held you close—but only to deduce

If what I felt was real or just a joke.

I whispered love—though merely to explain

Why my heart raced when you were standing near.

Each tender touch had reasons to maintain:

I kissed to test if love was truly here.

Explanatory Note: This poem demonstrates "pareuresis," a rhetorical figure that involves offering a seemingly reasonable excuse or pretext to justify an action, often one that the speaker actually wants to perform for entirely different reasons. The speaker provides rational-sounding explanations for each romantic action: kissing because the beloved's "lips were parted," holding close to "deduce" feelings, whispering love to "explain" physical reactions, and touching to "test" emotions. These pretexts create a veneer of logical justification for what are clearly acts of desire and affection. The figure allows the speaker to engage in romantic behavior while maintaining the fiction that each action serves some practical or investigative purpose rather than arising from simple emotional longing.

Kiss Bliss (with Pysma)

How can I know if kissing you was right?

I've asked the moon, but she just hides her face.

Should I have waited for a different night?

The stars won't answer, lost in endless space.

Was passion wrong, or was restraint the sin?

The wind sighs past but offers no reply.

Tell me, wise earth—where should I begin

To find the truth? But silence fills the sky.

Explanatory Note: This poem demonstrates "pysma," a rhetorical figure involving earnest inquiry or questioning, often directed toward nature, the divine, or abstract concepts, seeking guidance or answers about moral, philosophical, or personal dilemmas. The speaker addresses various elements of nature—the moon, stars, wind, and earth—asking fundamental questions about the rightness of their kiss and seeking wisdom about love and desire. Unlike simple rhetorical questions (which expect no answer), pysma represents genuine seeking and uncertainty, with the speaker truly hoping for guidance from these higher or external sources. The figure creates a tone of sincere perplexity and spiritual searching, though the silence that meets these inquiries suggests that some questions about love must be answered from within.

Kiss Bliss (with The Fallacy of Many Questions)

"When will you stop pretending that you care,

And why do you keep lying with each kiss?"

You ask, your eyes ablaze with cold despair,

As if my guilt were something I'd dismiss.

But wait—who said I'm lying or pretend?

These loaded questions trap me from the start.

I cannot answer what I won't defend:

False premises that tear my truth apart.

Explanatory Note: This poem demonstrates "the fallacy of many questions" (also called "complex question" or "loaded question"), which occurs when a question contains unproven assumptions that the respondent is forced to accept in order to provide any direct answer. The questions "When will you stop pretending that you care?" and "why do you keep lying with each kiss?" both assume that the speaker is indeed pretending and lying—assumptions that haven't been established as true. Any attempt to answer these questions directly would implicitly accept these false premises. The poem shows the speaker recognizing this logical trap and refusing to engage with questions that contain built-in accusations, instead challenging the unproven assumptions embedded within the interrogation itself.

Kiss Bliss (with Dissuasion)

Don't kiss me here beneath these watching stars—

Think of the scandal if we should be seen!

Consider how such passion always mars

The reputation of a maiden clean.

Remember all the hearts that kisses broke,

The promises that melted with the dawn.

Don't let desire become love's bitter yoke—

Walk away now, before all hope is gone.

Explanatory Note: This poem demonstrates "dissuasion," a rhetorical figure that involves systematically arguing against a particular course of action by presenting reasons, consequences, or warnings designed to discourage the listener from proceeding. The speaker uses multiple persuasive strategies to discourage kissing: appealing to fear of social judgment ("scandal if we should be seen"), invoking moral concerns ("reputation of a maiden"), citing historical precedent ("hearts that kisses broke"), and warning of future consequences ("promises that melted," "love's bitter yoke"). This creates a sustained argument against romantic action, though the very urgency and specificity of the warnings might paradoxically reveal the speaker's own conflicted desires about what they're ostensibly trying to prevent.

Kiss Bliss (with Persuasion)

Come, kiss me while the moonlight gilds the trees—

Life's fleeting moments beg us not to wait.

Consider how each heartbeat quickly flees,

How precious time makes all delay too late.

Think of the joy that one small kiss could bring,

The sweetness we might never know again.

Love's opportunities take sudden wing:

Kiss me now, before the night ends.

Explanatory Note: This poem demonstrates "persuasion," a rhetorical figure that involves systematically encouraging a particular course of action through appeals to emotion, logic, and urgency designed to move the listener toward a desired decision. The speaker employs multiple persuasive strategies: aesthetic appeal ("moonlight gilds the trees"), temporal urgency ("Life's fleeting moments," "each heartbeat quickly flees"), emotional reward ("joy that one small kiss could bring"), fear of loss ("sweetness we might never know again"), and metaphorical warning ("Love's opportunities take sudden wing"). This creates a sustained argument for immediate romantic action, building momentum through accumulated reasons that combine beauty, mortality, pleasure, and scarcity to overcome hesitation and encourage the beloved to act on desire.

Kiss Bliss (with Dehortation)

My dearest love, please do not kiss me now—

Though passion burns like fire within my chest.

Refrain from touching lips to fevered brow,

Though every nerve cries out to be caressed.

I beg you, hold back from that sweet embrace

That would undo us both with its sweet power.

Stay distant, love, though tears stream down my face:

Our kiss would bring destruction in this hour.

Explanatory Note: This poem demonstrates "dehortation," a rhetorical figure involving earnest advice or pleading against a particular course of action, often presented with genuine concern for consequences or well-being. Unlike simple dissuasion (which argues against something), dehortation carries a tone of urgent counsel from someone who cares about the person they're advising. The speaker uses tender address ("My dearest love") while desperately pleading against kissing ("please do not," "Refrain," "I beg you," "Stay distant"), acknowledging their own desire while warning of serious consequences ("would undo us both," "bring destruction"). This creates the emotional tension of someone torn between wanting something intensely and believing it would be harmful, making their counsel both heartfelt and conflicted.

Kiss Bliss (with Protrope)

Come, love, and kiss me while the roses bloom—
Embrace the joy that fate has placed before us!

Rise up from doubt and banish all your gloom:

Let passion's song ring out in sweetest chorus!

Take courage now and let your heart run free,

Cast off the chains of fear that hold you bound.

Move forward, darling, trust in you and me:

In kisses, love's true paradise is found.

Explanatory Note: This poem demonstrates "protrope," a rhetorical figure that involves encouragement, exhortation, or urgent advice to pursue a positive course of action, often using imperative language and motivational appeals. The speaker employs energetic commands ("Come," "Embrace," "Rise up," "Take courage," "Cast off," "Move forward") combined with inspiring imagery and reasons for action. Unlike simple persuasion, protrope has a more dynamic, rallying quality that seeks to rouse the listener from inaction or hesitation into bold engagement. The figure creates a tone of supportive urgency, like a coach encouraging an athlete or a friend boosting someone's confidence, using imperative verbs and uplifting language to inspire the beloved to overcome their reservations and act on their desires.

add to the wisdom of literary expression by writing a poem with 6 to 8 lines about a kiss. please be sure to use an example of the word schema or figure known as APOSIOPESIS in the poem and add the name of the figure or schema or trope (paranthetically) into the title like, "kiss bliss (with "the name of the figure')" also can you put and explanatory note at the bottom of the poem? and the title at the beginning

Edit
Kiss Poem with Aposiopesis
Document
Kiss Bliss (with Aposiopesis) Your lips approach mine in the golden light, Time slows to honey, thick and sweet— But then... the moment breaks apart, Words flee like startled birds at dawn. What follows after cannot be spoken, Only felt in the tre

I've crafted a poem that captures the ineffable moment of a kiss through the deliberate use of aposiopesis. The breaking off of the phrase "But then..." mirrors how overwhelming intimacy can render us speechless, creating that pregnant pause where feeling transcends language. The device works particularly well in poetry about intimate moments, as it invites the reader into that space of unspoken experience.

Retry
N

add to the wisdom of literary expression by writing a poem with 6 to 8 lines about a kiss. please be sure to use an example of the word schema or figure known as ECPHONESIS in the poem and add the name of the figure or schema or trope (paranthetically) into the title like, "kiss bliss (with "the name of the figure')" also can you put and explanatory note at the bottom of the poem? and the title at the beginning

Edit

Kiss Poem with Ecphonesis

Document · Version 2

Kiss Bliss (with Ecphonesis) Your lips find mine beneath the silver moon, And oh! what lightning strikes between our souls— The world dissolves in that electric touch, Time bends and breaks around us like glass. Ah, sweet surrender! How the heart

I've rewritten the poem to feature ecphonesis, which perfectly captures the spontaneous emotional eruption that accompanies profound intimacy. The exclamatory outbursts "And oh!" and "Ah, sweet surrender!" give voice to those moments when feeling overwhelms our capacity for measured speech, creating peaks of intensity that mirror the physical and emotional crescendo of a kiss.

Retry

N

add to the wisdom of literary expression by writing a poem with 6 to 8 lines about a kiss. please be sure to use an example of the word schema or figure known as THAUMASMUS in the poem and add the name of the figure or schema or trope (paranthetically) into the title like, "kiss bliss (with "the name of the figure')" also can you put and explanatory note at the bottom of the poem? and the title at the beginning

Edit

Kiss Poem with Thaumasmus

Document · Version 3

Kiss Bliss (with Thaumasmus) Your lips approach mine in the fading light, And I wonder—how can such simple flesh Hold galaxies? How does this gentle touch Unravel all the careful walls I've built? What strange alchemy transforms two souls Into one

I've crafted a poem using thaumasmus to capture the wonder and bewilderment that accompanies intimate connection. The series of amazed questions—about how simple flesh can hold galaxies, what alchemy transforms souls, and how time bows to love—creates a sense of reverent questioning that mirrors how profound experiences often leave us marveling at their impossible beauty and power.

Retry

N

add to the wisdom of literary expression by writing a poem with 6 to 8 lines about a kiss. please be sure to use an example of the word schema or figure known as PEROTEMA in the poem and add the name of the figure or schema or trope (paranthetically) into the title like, "kiss bliss (with "the name of the figure')" also can you put and explanatory note at the bottom of the poem? and the title at the beginning

Edit
Kiss Poem with Perotema
Document · Version 4
Kiss Bliss (with Perotema) Your lips meet mine beneath the starlit sky, And in that moment—who could doubt such grace? The world stands still, the very air grows thick With sweetness that no honey ever held. What heart could remain unmoved by this

I've created a poem employing perotema to emphasize the undeniable power of an intimate kiss. The rhetorical questions—"who could doubt such grace?" and "What heart could remain unmoved?"—invite the reader to acknowledge what we all know to be true: that love's touch is universally moving and impossible to resist. This device strengthens the poem's emotional argument by presenting the kiss's power as beyond question or debate.

Retry
N

add to the wisdom of literary expression by writing a poem with 6 to 8 lines about a kiss. please be sure to use an example of the word schema or figure known as EROTEMA in the poem and add the name of the figure or schema or trope (paranthetically) into the title like, "kiss bliss (with "the name of the figure')" also can you put and explanatory note at the bottom of the poem? and the title at the beginning

Edit
Kiss Poem with Erotema
Document · Version 5
Kiss Bliss (with Erotema) Your lips approach mine in the morning light, But should I dare to close this sacred space? The moment trembles on desire's edge, My heart beats wild with longing's fierce demand. Will you think me too bold if I lean clos

I've written a poem using erotema to capture the genuine uncertainty and vulnerability that often precedes intimate moments. The authentic questions—"should I dare?" and "Will you think me too bold?"—reflect the real hesitation and tender anxiety that can accompany the approach to a kiss, making the poem more emotionally honest and relatable than rhetorical questions with obvious answers would be.

Retry
N

add to the wisdom of literary expression by writing a poem with 6 to 8 lines about a kiss. please be sure to use an example of the word schema or figure known as APOSTROPHE in the poem and add the name of the figure or schema or trope (paranthetically) into the title like, "kiss bliss (with "the name of the figure')" also can you put and explanatory note at the bottom of the poem? and the title at the beginning

Edit
Kiss Poem with Apostrophe
Document · Version 6
Kiss Bliss (with Apostrophe) Your lips find mine beneath the evening stars, O Time, stand still and hold this moment fast! The world dissolves in honey-sweet embrace, While passion's fire burns through my trembling soul. Sweet Love, you cruel and

I've crafted a poem using apostrophe to address abstract forces—Time and Love—as if they were present beings who could hear the speaker's pleas. This device heightens the drama and emotional intensity of the kiss by making the speaker's inner experience feel cosmic in scope, as if the entire universe were witness to and participant in this intimate moment.

Retry
N

add to the wisdom of literary expression by writing a poem with 6 to 8 lines about a kiss. please be sure to use an example of the word schema or figure known as ANAMNESIS in the poem and add the name of the figure or schema or trope (paranthetically) into the title like, "kiss bliss (with "the name of the figure')" also can you put and explanatory note at the bottom of the poem? and the title at the beginning

Edit
Kiss Poem with Anamnesis
Document · Version 7

Kiss Bliss (with Anamnesis) Your lips touch mine in twilight's purple glow, And suddenly I remember—yes, remember now— How the first lovers felt when time began, That primal fire that sparked the world to life. This kiss awakens something ancient,

I've written a poem employing anamnesis to suggest that a kiss awakens deep, ancestral memories within us. The phrase "And suddenly I remember—yes, remember now—" captures that moment when intimate experience seems to unlock something ancient and universal, as if each kiss connects us to the very first expressions of human love and desire throughout history.

Retry
N

add to the wisdom of literary expression by writing a poem with 6 to 8 lines about a kiss. please be sure to use an example of the word schema or figure known as APOCARTERESIS in the poem and add the name of the figure or schema or trope (paranthetically) into the title like, "kiss bliss (with "the name of the figure')" also can you put and explanatory note at the bottom of the poem? and the title at the beginning

Edit
Kiss Poem with Apocarteresis
Document · Version 8
Kiss Bliss (with Apocarteresis) Your lips approach mine in the candlelit room, And I should pull away—yes, I should resist This dangerous sweetness that calls to

me. I ought to guard my heart from such surrender, But oh, how can I turn from paradi

I've crafted a poem using apocarteresis to capture that moment when we recognize we should resist something dangerous or overwhelming, yet find ourselves powerless to do so. The speaker acknowledges they "should pull away" and "ought to guard" their heart, but ultimately surrenders to the kiss with resigned acceptance of love's irresistible force—a beautiful tension between wisdom and desire.

Retry
N

add to the wisdom of literary expression by writing a poem with 6 to 8 lines about a kiss. please be sure to use an example of the word schema or figure known as OPTATIO in the poem and add the name of the figure or schema or trope (paranthetically) into the title like, "kiss bliss (with "the name of the figure')" also can you put and explanatory note at the bottom of the poem? and the title at the beginning

Edit
Kiss Poem with Optatio
Document · Version 9
Kiss Bliss (with Optatio) Your lips hover near mine in morning's light, Would that this moment could stretch into years! The space between us trembles with desire, Oh, would that time itself might stop its flight And grant us this eternity of touc

I've written a poem employing optatio to express the speaker's passionate desire to make the moment of the kiss eternal. The wishful phrases "Would that this moment could stretch into years!" and "If only dawn would never break this spell" capture that universal longing to freeze perfect moments in time, giving voice to the way profound intimacy

Kiss Bliss (with Aposiopesis)

Your lips approach mine in the golden light,

Time slows to honey, thick and sweet—

But then... the moment breaks apart,

Words flee like startled birds at dawn.

What follows after cannot be spoken,

Only felt in the trembling space

Between your breath and mine.

Note: Aposiopesis is a rhetorical device where a sentence is deliberately left unfinished, creating a dramatic pause or suggesting that the speaker is too overcome by emotion to continue. In this poem, "But then..." demonstrates this figure, as the speaker becomes too overwhelmed by the intimacy of the moment to complete their thought, leaving the reader to imagine what transcends language itself.

Kiss Bliss (with Ecphonesis)

Your lips find mine beneath the silver moon,

And oh! what lightning strikes between our souls—

The world dissolves in that electric touch,

Time bends and breaks around us like glass.

Ah, sweet surrender! How the heart forgets

To beat in measured time, instead it soars

On wings of pure and breathless wonder.

Note: Ecphonesis is a rhetorical figure involving an exclamatory phrase or emotional outburst, often beginning with "O!" or "Ah!" In this poem, "And oh!" and "Ah, sweet surrender!" demonstrate this device, expressing the overwhelming emotion and wonder that accompanies an intimate kiss—feelings too intense for ordinary declarative language.

Kiss Bliss (with Thaumasmus)

Your lips approach mine in the fading light,

And I wonder—how can such simple flesh

Hold galaxies? How does this gentle touch

Unravel all the careful walls I've built?

What strange alchemy transforms two souls

Into one burning constellation? How

Does time itself bow down to love's brief fire?

Note: Thaumasmus is a rhetorical figure expressing wonder or amazement through a series of questions, often highlighting the miraculous or inexplicable nature of something. In this poem, the cascading questions—"how can such simple flesh / Hold galaxies?" and "What strange alchemy transforms two souls"—capture the speaker's awe at love's transformative power, emphasizing how a kiss transcends physical reality to become something mystical and wondrous.

Kiss Bliss (with Perotema)

Your lips meet mine beneath the starlit sky,

And in that moment—who could doubt such grace?

The world stands still, the very air grows thick

With sweetness that no honey ever held.

What heart could remain unmoved by this?

Time bends around us like a gentle wave,

Who would not surrender to such bliss?

Note: Perotema is a rhetorical figure that poses questions with obvious answers, often used to emphasize a point through seeming inquiry. In this poem, "who could doubt such grace?" "What heart could remain unmoved by this?" and "Who would not surrender to such bliss?" are questions whose answers are self-evident—no one could doubt, no heart could remain unmoved, everyone would surrender. This device reinforces the universal and undeniable power of love's kiss.

Kiss Bliss (with Erotema)

Your lips approach mine in the morning light,

But should I dare to close this sacred space?

The moment trembles on desire's edge,

My heart beats wild with longing's fierce demand.

Will you think me too bold if I lean close?

Or shall I wait for you to bridge the gap

Between what is and what we both desire?

Note: Erotema is a rhetorical figure involving genuine questions that seek real answers or express doubt and uncertainty. Unlike perotema (which poses questions with obvious answers), erotema presents authentic inquiry. In this poem, "But should I dare to close this sacred space?" and "Will you think me too bold if I lean close?" express the speaker's genuine uncertainty and vulnerability in the intimate moment, capturing the real hesitation and questioning that often accompanies tender encounters.

Kiss Bliss (with Apostrophe)

Your lips find mine beneath the evening stars,

O Time, stand still and hold this moment fast!

The world dissolves in honey-sweet embrace,

While passion's fire burns through my trembling soul.

Sweet Love, you cruel and tender master,

How you bind us with invisible chains

Of longing that no earthly force can break.

Note: Apostrophe is a rhetorical figure in which the speaker directly addresses someone or something that is not present, dead, or abstract. In this poem, the speaker addresses "Time" ("O Time, stand still") and "Love" ("Sweet Love, you cruel and tender master"), calling upon these abstract concepts as if they were present entities capable of hearing and responding. This device intensifies the emotional urgency of the moment and gives voice to the speaker's desire to transcend ordinary reality.

Kiss Bliss (with Anamnesis)

Your lips touch mine in twilight's purple glow,

And suddenly I remember—yes, remember now—

How the first lovers felt when time began,

That primal fire that sparked the world to life.

This kiss awakens something ancient, deep,

A memory older than my beating heart,

The very blueprint of desire itself.

Note: Anamnesis is a rhetorical figure involving the deliberate recalling or recovering of memory, often presenting past events as if suddenly remembered or revealed. In this poem, "And suddenly I remember—yes, remember now—" demonstrates this device, as the speaker experiences the kiss as unlocking ancient, almost primordial memories of love's first expression, suggesting that intimate connection taps into something timeless and universal within human experience.

Kiss Bliss (with Apocarteresis)

Your lips approach mine in the candlelit room,

And I should pull away—yes, I should resist

This dangerous sweetness that calls to me.

I ought to guard my heart from such surrender,

But oh, how can I turn from paradise?

Your kiss dissolves my will like morning frost,

And I am lost, gloriously, completely lost.

Note: Apocarteresis is a rhetorical figure expressing patient endurance or resignation in the face of something one feels powerless to resist or change. In this poem, "And I should pull away—yes, I should resist" and "I ought to guard my heart" demonstrate this device, as the speaker acknowledges what they should do to protect themselves, yet ultimately accepts their inability to resist love's overwhelming power, surrendering with a kind of resigned joy to the inevitable.

Kiss Bliss (with Optatio)

Your lips hover near mine in morning's light,

Would that this moment could stretch into years!

The space between us trembles with desire,

Oh, would that time itself might stop its flight

And grant us this eternity of touch.

If only dawn would never break this spell,

We'd live forever in love's perfect kiss.

Note: Optatio is a rhetorical figure expressing a passionate wish or desire, often introduced by phrases like "would that," "if only," or "oh that." In this poem, "Would that this moment could stretch into years!" "Oh, would that time itself might stop its flight" and "If only dawn would never break this spell" demonstrate this device, as the speaker expresses fervent wishes for the kiss and its accompanying moment to transcend temporal limitations and become eternal.

add to the wisdom of literary expression by writing a poem with 6 to 8 lines about a kiss. please be sure to use an example of the word schema or figure known as OPTATIO in the poem and add the name of the figure or schema or trope (paranthetically) into the title like, "kiss bliss (with "the name of the figure')" also can you put and explanatory note at the bottom of the poem? and the title at the beginning

Kiss Poem with Optatio

Document

I've crafted a poem that captures the intensity of a romantic moment through the lens of optatio, where the speaker's passionate wishes transform a simple kiss into something transcendent. The rhetorical figure creates a sense of yearning that elevates the physical moment into the realm of the eternal and sublime.

add to the wisdom of literary expression by writing a poem with 6 to 8 lines about a kiss. please be sure to use an example of the word schema or figure known as SUPPLICATION in the poem and add the name of the figure or schema or trope (paranthetically) into the title like, "kiss bliss (with "the name of the figure')" also can you put and explanatory note at the bottom of the poem? and the title at the beginning

Kiss Poem with Supplication

Document · Version 2

I've created a new poem that employs supplication to transform the desire for a kiss into a humble petition. The speaker adopts the voice of a reverent supplicant, using pleading language that turns romantic longing into an act of devotional entreaty, creating both vulnerability and dignity in the request.

add to the wisdom of literary expression by writing a poem with 6 to 8 lines about a kiss. please be sure to use an example of the word schema or figure known as MEMPSIS in the poem and add the name of the figure or schema or trope (paranthetically) into the title like, "kiss bliss (with "the name of the figure')" also can you put and explanatory note at the bottom of the poem? and the title at the beginning

Kiss Poem with Mempsis

Document · Version 3

I've crafted a poem that uses mempsis to create dramatic tension around a kiss. The speaker employs this rhetorical figure of complaint and counter-accusation to defend their romantic action, turning the tables on an implied critic by highlighting the contradictions in their blame. This creates a more complex emotional landscape where desire, guilt, and justification intersect.

add to the wisdom of literary expression by writing a poem with 6 to 8 lines about a kiss. please be sure to use an example of the word schema or figure known as PARAMYTHIA in the poem and add the name of the figure or schema or trope (paranthetically) into the title like, "kiss bliss (with "the name of the figure')" also can you put and explanatory note at the bottom of the poem? and the title at the beginning

Kiss Poem with Paramythia

Document · Version 4

I've created a poem that uses paramythia to offer gentle consolation about a lost or distant kiss. The speaker assumes the role of a compassionate counselor, using soothing language and hopeful perspectives to comfort someone experiencing romantic grief. This rhetorical figure transforms the poem into an act of emotional healing, showing how love's memory can provide solace even in separation.

add to the wisdom of literary expression by writing a poem with 6 to 8 lines about a kiss. please be sure to use an example of the word schema or figure known as MEDELA in the poem and add the name of the figure or schema or trope (paranthetically) into the title like, "kiss bliss (with "the name of the figure')" also can you put and explanatory note at the bottom of the poem? and the title at the beginning

Kiss Poem with Medela

Document · Version 5

I've crafted a poem that uses medela to transform a kiss into a powerful healing metaphor. The speaker presents the kiss as medicine, cure, and remedy, creating a sustained metaphorical framework where romantic love becomes a therapeutic force capable of healing emotional wounds and restoring wholeness. This rhetorical figure elevates the simple act of kissing into something with transformative, almost miraculous healing properties.

add to the wisdom of literary expression by writing a poem with 6 to 8 lines about a kiss. please be sure to use an example of the word schema or figure known

as MENEMIUS in the poem and add the name of the figure or schema or trope (paranthetically) into the title like, "kiss bliss (with "the name of the figure')" also can you put and explanatory note at the bottom of the poem? and the title at the beginning

Menemius rhetorical figure definition classical rhetoric

10 results

"Menemius" classical rhetoric figure definition

0 results

I'm having difficulty finding a specific definition for "Menemius" as a classical rhetorical figure in the search results. This term doesn't appear to be a widely documented or recognized classical rhetorical figure in the standard sources I can access.

It's possible that:

1. This might be a very obscure or specialized term

2. There could be a spelling variation or transliteration issue
3. It might be from a specific textual tradition not covered in general rhetoric sources

Could you provide more context about where you encountered "Menemius" as a rhetorical figure, or confirm the spelling? Alternatively, would you like me to work with a different classical rhetorical figure that I can definitively identify and explain? I'd be happy to create a poem using any of the well-documented figures like apostrophe, prosopopoeia, or others from the classical tradition.

add to the wisdom of literary expression by writing a poem with 6 to 8 lines about a kiss. please be sure to use an example of the word schema or figure known as MENENIUS in the poem and add the name of the figure or schema or trope (paranthetically) into the title like, "kiss bliss (with "the name of the figure')" also can you put and explanatory note at the bottom of the poem? and the title at the beginning

Menenius rhetorical figure classical rhetoric definition

10 results

W

W

W

W

"Menenius" classical rhetorical figure definition ancient rhetoric

0 results

I apologize, but I'm unable to find a specific definition for "Menenius" as a classical rhetorical figure in the sources I can access. This term doesn't appear in the standard glossaries and references for classical rhetoric that I've searched.

It's possible that:

1. This might be a very specialized or obscure term from a particular rhetorical tradition
2. There could be a variant spelling or transliteration
3. It might be from a specific ancient text or commentary not widely catalogued

Would you be able to provide more context about where you encountered "Menenius" as a rhetorical figure? Or would you prefer I work with a different, well-documented classical rhetorical figure like apostrophe, epanalepsis, or chiasmus? I'd be happy to create a compelling poem about a kiss using any of the established figures from the classical tradition.

add to the wisdom of literary expression by writing a poem with 6 to 8 lines about a kiss. please be sure to use an example of the word schema or figure known as MYCTERISMUS in the poem and add the name of the figure or schema or trope (paranthetically) into the title like, "kiss bliss (with "the name of the figure')" also can you put and explanatory note at the bottom of the poem? and the title at the beginning

Kiss Poem with Mycterismus

Document · Version 6

I've created a poem that uses mycterismus to transform romantic disappointment into biting satirical commentary. The speaker employs this figure of scornful mockery to ironically "praise" a cold, reluctant kiss, using sarcastic language that creates a sharp contrast between expectation and reality. This rhetorical figure adds a sophisticated layer of bitter wit to the exploration of romantic experience.

add to the wisdom of literary expression by writing a poem with 6 to 8 lines about a kiss. please be sure to use an example of the word schema or figure known as SARCASMUS in the poem and add the name of the figure or schema or trope (paranthetically) into the title like, "kiss bliss (with "the name of the figure')" also can you put and explanatory note at the bottom of the poem? and the title at the beginning

Kiss Poem with Sarcasmus

Document · Version 7

I've crafted a poem that employs sarcasmus to express the pain of romantic betrayal through bitter irony. The speaker uses seemingly positive language—"wonderful," "thrilling," "tender"—to actually convey deep hurt and contempt for a deceptive kiss. This rhetorical figure creates a powerful tension between the surface meaning and the underlying emotional reality, transforming romantic disillusionment into sharp, cutting commentary.

add to the wisdom of literary expression by writing a poem with 6 to 8 lines about a kiss. please be sure to use an example of the word schema or figure known as EPIPLEXIS in the poem and add the name of the figure or schema or trope (paranthetically) into the title like, "kiss bliss (with "the name of the figure')" also can you put and explanatory note at the bottom of the poem? and the title at the beginning

Kiss Poem with Epiplexis

Document · Version 8

I've created a poem that uses epiplexis to express the emotional confusion and hurt that comes from mixed romantic signals. The speaker employs this figure of reproachful questioning to confront their lover about the contradictions between intimate actions and distant behavior, transforming romantic bewilderment into a series of pointed, emotional questions that reveal the depth of their confusion and pain.

add to the wisdom of literary expression by writing a poem with 6 to 8 lines about a kiss. please be sure to use an example of the word schema or figure known as PERCONTATIO in the poem and add the name of the figure or schema or trope (paranthetically) into the title like, "kiss bliss (with "the name of the figure')" also can you put and explanatory note at the bottom of the poem? and the title at the beginning

Kiss Poem with Percontatio

Document · Version 9

I've crafted a poem that uses percontatio to transform a kiss into a subject of philosophical wonder and inquiry. The speaker employs this figure of earnest questioning to explore the mysterious power of romantic touch, using genuine curiosity to investigate how a simple kiss can have such profound, seemingly magical effects. This creates a sense of awe and reverence around the romantic moment through sincere intellectual exploration.

add to the wisdom of literary expression by writing a poem with 6 to 8 lines about a kiss. please be sure to use an example of the word schema or figure known as ONEDISMUS in the poem and add the name of the figure or schema or trope (paranthetically) into the title like, "kiss bliss (with "the name of the figure')" also can you put and explanatory note at the bottom of the poem? and the title at the beginning

Onedismus rhetorical figure classical rhetoric definition

10 results

W

W

ufr

UK

W

W

"Onedismus" classical rhetoric figure definition ancient rhetoric

0 results

I apologize, but I'm unable to find a specific definition for "Onedismus" as a classical rhetorical figure in the sources I can access. This term doesn't appear in the standard glossaries and references for classical rhetoric that I've searched, including comprehensive lists of Greek and Latin rhetorical terms.

It's possible that:

1. This might be a very specialized or obscure term from a particular rhetorical tradition
2. There could be a variant spelling or transliteration issue
3. It might be from a specific ancient text or commentary not widely catalogued in modern sources

Would you be able to provide more context about where you encountered "Onedismus" as a rhetorical figure? Or would you prefer I work with a different, well-documented classical rhetorical figure like apostrophe, litotes, or anadiplosis? I'd be happy to create a compelling poem about a kiss using any of the established figures from the classical tradition.

add to the wisdom of literary expression by writing a poem with 6 to 8 lines about a kiss. please be sure to

use an example of the word schema or figure known as CATEGORIA in the poem and add the name of the figure or schema or trope (paranthetically) into the title like, "kiss bliss (with "the name of the figure')" also can you put and explanatory note at the bottom of the poem? and the title at the beginning

Kiss Poem with Categoria

Document · Version 10

I've created a poem that uses categoria to transform a romantic encounter into a playful mock trial. The speaker employs this figure of direct accusation to formally "charge" their lover with various romantic "crimes," using legal terminology to create a witty framework where being kissed becomes a delightful offense. This rhetorical approach adds both humor and intensity to the exploration of romantic power dynamics.

add to the wisdom of literary expression by writing a poem with 6 to 8 lines about a kiss. please be sure to use an example of the word schema or figure known as PROCLEES in the poem and add the name of the figure or schema or trope (paranthetically) into the title like, "kiss bliss (with "the name of the figure')" also can you put and explanatory note at the bottom of the poem? and the title at the beginning

Proclees rhetorical figure classical rhetoric definition

10 results

W

P•

Fetched proclees

byu.edu

Based on my search, I found that proclees is a rhetorical device that emphasizes a point by framing it positively Rhetoric - Wikipedia and relates to challenging one's adversary What is Classical Rhetoric? - Memoria Press: Classical Education. This suggests proclees is a figure that involves issuing a challenge or provocation while maintaining a positive framing.

Kiss Poem with Proclees

Document · Version 11

I've created a poem that uses proclees to transform romantic desire into a playful challenge. The speaker employs this figure to dare their lover to prove their romantic abilities through their kiss, using positive framing to create an enticing provocation. This rhetorical approach adds both tension and invitation to the romantic encounter, making the kiss both a test and a reward.

add to the wisdom of literary expression by writing a poem with 6 to 8 lines about a kiss. please be sure to

use an example of the word schema or figure known as BDELYGMIA in the poem and add the name of the figure or schema or trope (paranthetically) into the title like, "kiss bliss (with "the name of the figure')" also can you put and explanatory note at the bottom of the poem? and the title at the beginning

Kiss Poem with Bdelygmia

Document · Version 12

I've created a poem that uses bdelygmia to explore the complex, contradictory emotions that can arise from a powerful romantic encounter. The speaker employs this figure of disgust and loathing to express their conflicted feelings about a kiss that has left them both devastated and yearning, creating a fascinating paradox where revulsion and desire intertwine in the aftermath of intense romantic connection.

add to the wisdom of literary expression by writing a poem with 6 to 8 lines about a kiss. please be sure to use an example of the word schema or figure known as CATAPLEXIS in the poem and add the name of the figure or schema or trope (paranthetically) into the title like, "kiss bliss (with "the name of the figure')" also can you put and explanatory note at the bottom of the poem? and the title at the beginning

Kiss Poem with Cataplexis

Document · Version 13

I've crafted a poem that uses cataplexis to capture the overwhelming impact of an unexpected kiss. The speaker employs this figure of amazement and astonishment to convey how the romantic encounter leaves them completely stunned and disoriented, using language that emphasizes the shocking power of love to transform and overwhelm in a single moment of intimate contact.

add to the wisdom of literary expression by writing a poem with 6 to 8 lines about a kiss. please be sure to use an example of the word schema or figure known as ARA in the poem and add the name of the figure or schema or trope (paranthetically) into the title like, "kiss bliss (with "the name of the figure')" also can you put and explanatory note at the bottom of the poem? and the title at the beginning

Kiss Poem with Ara

Document · Version 14

I've created a poem that uses ara to transform romantic betrayal into a formal curse. The speaker employs this figure of malediction to call down punishment upon the kiss and lover that deceived them, using ritualistic cursing language that elevates personal hurt into the realm of divine justice. This rhetorical approach adds both gravity and cathartic power to the expression of romantic disappointment.

add to the wisdom of literary expression by writing a poem with 6 to 8 lines about a kiss. please be sure to use an example of the word schema or figure known as CURSING in the poem and add the name of the figure or schema or trope (paranthetically) into the title like, "kiss bliss (with "the name of the figure')" also can you put and explanatory note at the bottom of the poem? and the title at the beginning

Kiss Poem with Cursing

Document · Version 15

I've crafted a poem that uses cursing to express the intense, conflicted emotions of overwhelming romantic passion. The speaker employs profane language to simultaneously condemn and cherish a transformative kiss, creating a powerful paradox where they curse the very experience that has moved them most deeply. This rhetorical approach captures the raw intensity of love's ability to both elevate and torment the human heart.

add to the wisdom of literary expression by writing a poem with 6 to 8 lines about a kiss. please be sure to use an example of the word schema or figure known as EULOGIA in the poem and add the name of the figure or schema or trope (paranthetically) into the title like, "kiss bliss (with "the name of the figure')" also can you put and explanatory note at the bottom of the poem? and the title at the beginning

Kiss Poem with Eulogia

Document · Version 16

I've created a poem that uses eulogia to transform a romantic kiss into a sacred blessing. The speaker employs this figure of praise and reverent speech to elevate the intimate encounter to spiritual dimensions, using liturgical language that treats the kiss as a form of divine blessing. This rhetorical approach creates a sense of reverence and sanctity around the romantic moment, suggesting that true love partakes of the sacred.

add to the wisdom of literary expression by writing a poem with 6 to 8 lines about a kiss. please be sure to use an example of the word schema or figure known as PAEANISMUS in the poem and add the name of the figure or schema or trope (paranthetically) into the title like, "kiss bliss (with "the name of the figure')" also can you put and explanatory note at the bottom of the poem? and the title at the beginning

Paeanismus rhetorical figure classical rhetoric definition

10 results

"Paeanismus" classical rhetoric figure definition triumph song praise

0 results

Based on my understanding of classical rhetoric, "paeanismus" would be related to the paean, which was a song of praise or triumph in ancient Greek culture, often sung in honor of Apollo or to celebrate victory. As a rhetorical figure, paeanismus would involve expressing triumph, celebration, or exuberant praise.

Kiss Poem with Paeanismus

Document · Version 17

I've created a poem that uses paeanismus to transform a romantic kiss into a moment of triumphant celebration. The speaker employs this figure of exuberant praise to celebrate the kiss as a glorious victory over doubt and loneliness, using language of triumph and jubilation that elevates the intimate moment to the level of a public celebration worthy of song and proclamation.

add to the wisdom of literary expression by writing a poem with 6 to 8 lines about a kiss. please be sure to use an example of the word schema or figure known as COMPROBATIO in the poem and add the name of the figure or schema or trope (paranthetically) into the title like, "kiss bliss (with "the name of the figure')" also can you put and explanatory note at the bottom of the poem? and the title at the beginning

Kiss Poem with Comprobatio

Document · Version 18

I've crafted a poem that uses comprobatio to validate and confirm the transformative power of a kiss. The speaker employs this figure of acknowledgment and testimony to bear witness to love's undeniable effects, using confirmatory language that treats the romantic experience as an established truth that must be recognized and affirmed. This rhetorical approach adds credibility and solemnity to the celebration of romantic intimacy.

add to the wisdom of literary expression by writing a poem with 6 to 8 lines about a kiss. please be sure to use an example of the word schema or figure known as PARRHESIA in the poem and add the name of the figure or schema or trope (paranthetically) into the title like, "kiss bliss (with "the name of the figure')" also can you put and explanatory note at the bottom of the poem? and the title at the beginning

Kiss Poem with Parrhesia

Document · Version 19

I've created a poem that uses parrhesia to confront the difficult truth about romantic authenticity. The speaker employs this figure of fearless, frank speech to address the painful possibility that a kiss might be empty of genuine feeling, choosing brutal honesty over comfortable illusion. This rhetorical approach adds moral courage to the exploration of love, valuing truth over pleasantness in matters of the heart.

add to the wisdom of literary expression by writing a poem with 6 to 8 lines about a kiss. please be sure to use an example of the word schema or figure known as EUCHARISTIA in the poem and add the name of the figure or schema or trope (paranthetically) into the title like, "kiss bliss (with "the name of the figure')" also can you put and explanatory note at the bottom of the poem? and the title at the beginning

Kiss Poem with Eucharistia

Document · Version 20

I've crafted a poem that uses eucharistia to transform a romantic kiss into an occasion for profound gratitude. The speaker employs this figure of thanksgiving to express appreciation for love as a gift, using grateful language that treats the romantic encounter as a blessing worthy of acknowledgment and thanks. This rhetorical approach adds humility and reverence to the celebration of intimate connection.

add to the wisdom of literary expression by writing a poem with 6 to 8 lines about a kiss. please be sure to use an example of the word schema or figure known as SYNGNOME in the poem and add the name of the figure or schema or trope (paranthetically) into the title like, "kiss bliss (with "the name of the figure')" also can you put and explanatory note at the bottom of the poem? and the title at the beginning

Kiss Poem with Syngnome

Document · Version 21

I've created a poem that uses syngnome to seek understanding and forgiveness for the overwhelming emotional response to a powerful kiss. The speaker employs this figure of pardon to excuse their intense reactions, appealing to the universal nature of love's power to overwhelm even the most composed hearts. This rhetorical approach adds vulnerability and humility to the expression of romantic intensity.

add to the wisdom of literary expression by writing a poem with 6 to 8 lines about a kiss. please be sure to use an example of the word schema or figure known

as EXCLAMATIO in the poem and add the name of the figure or schema or trope (paranthetically) into the title like, "kiss bliss (with "the name of the figure')" also can you put and explanatory note at the bottom of the poem? and the title at the beginning

Kiss Poem with Exclamatio

Document · Version 22

I've crafted a poem that uses exclamatio to express the overwhelming emotional intensity of a romantic kiss through passionate outcries. The speaker employs this figure of emotional exclamation to create a cascade of feelings that cannot be contained within normal speech, using interjections and exclamatory phrases to capture the transformative power of love's touch. This rhetorical approach creates immediacy and raw emotional impact.

add to the wisdom of literary expression by writing a poem with 6 to 8 lines about a kiss. please be sure to use an example of the word schema or figure known as EXUSCITATIO in the poem and add the name of the figure or schema or trope (paranthetically) into the title like, "kiss bliss (with "the name of the figure')" also can you put and explanatory note at the bottom of the poem? and the title at the beginning

Kiss Poem with Exuscitatio

Document · Version 23

I've created a poem that uses exuscitatio to capture the awakening power of a transformative kiss. The speaker employs this figure of arousal and awakening to stir their own dormant romantic feelings, using imperative language to rouse heart, soul, and spirit to embrace love's awakening force. This rhetorical approach creates a sense of spiritual and emotional resurrection through the power of romantic connection.

add to the wisdom of literary expression by writing a poem with 6 to 8 lines about a kiss. please be sure to use an example of the word schema or figure known as PARALIPSIS in the poem and add the name of the figure or schema or trope (paranthetically) into the title like, "kiss bliss (with "the name of the figure')" also can you put and explanatory note at the bottom of the poem? and the title at the beginning

Kiss Poem with Paralipsis

Document · Version 24

I've crafted a poem that uses paralipsis to reveal the profound impact of a kiss while pretending to avoid discussing it. The speaker employs this figure of deliberate omission to actually emphasize the very romantic experiences they claim not to mention, creating an ironic tension where the act of "not speaking" becomes the most eloquent way to communicate the kiss's transformative power. This rhetorical approach adds sophistication and playful subtlety to the expression of intimate feeling.

Kiss Bliss (with Optatio)

Would that this moment could stretch beyond time,

Your lips pressed soft against mine in the dark—

Oh, let the world forget to turn and chime,

That we might dwell within love's sacred spark.

If only breath itself would pause its flight,

And hold us here in this eternal light,

Where nothing exists but your kiss and mine,

Sweet願望 wrapped in rhythm and rhyme.

Explanatory Note: Optatio is a rhetorical figure expressing an ardent wish or desire, typically introduced by phrases like "would that," "oh that," "if only," or "let." In this poem, the speaker employs optatio to express their fervent desire for the kiss and the moment to transcend the ordinary limits of time, using the wishful constructions "Would that," "Oh, let," and "If only" to convey the intensity of their longing for the romantic moment to become eternal.

Kiss Bliss (with Supplication)

I beg you, let your lips find mine tonight,

Grant me this mercy in the fading light—

Please, do not turn away or hesitate,

I supplicate your kiss to seal my fate.

Have pity on this heart that burns for you,

Bestow upon me love both sweet and true,

For I am but a beggar at your door,

Pleading for the kiss I'm longing for.

Explanatory Note: Supplication is a rhetorical figure that involves earnest entreaty or humble petition, often expressed through pleading language such as "I beg," "please," "have pity," "grant me," and "I supplicate." In this poem, the speaker assumes the posture of a humble petitioner, using supplicatory language to transform the request for a kiss into an act of devotion and vulnerability, elevating the romantic gesture through the dignity of earnest pleading.

Kiss Bliss (with Mempsis)

How dare you blame me for this stolen kiss,

When your own eyes invited my advance?

You chide me for the moment's gentle bliss,

Yet gave me every signal, every glance.

Why scold the fire when you struck the match?

Your lips, not mine, first whispered "stay and catch

This fleeting breath between us in the night"—

So mock not what your heart made feel so right.

Explanatory Note: Mempsis is a rhetorical figure involving complaint or accusation, often used to counter-blame or express indignation at being wrongly accused. In this poem, the speaker employs mempsis to defend against an implied accusation about the kiss, turning the blame back on their partner by pointing out the contradictions in their complaint—how they invited the very action they now criticize, creating dramatic tension through this defensive counter-accusation.

Kiss Bliss (with Paramythia)

Do not weep, beloved, for that first kiss lost—

Time heals all wounds, and love will come again.
Though distance now divides what passion crossed,

Take comfort: hearts remember more than pain.

Be not dismayed by separation's sting,

For every ending births a new beginning.

Consider how the sweetest memories stay

To light the darkness of your lonely day.

Explanatory Note: Paramythia is a rhetorical figure of consolation or comfort, used to soothe someone's grief or distress through gentle encouragement and reassuring words. In this poem, the speaker employs paramythia to console someone mourning a lost kiss or separated love, offering comfort through reminders that time heals, love endures in memory, and hope remains for the future—transforming sorrow into solace through compassionate counsel.

Kiss Bliss (with Medela)

Your kiss became the cure for all my pain,

A healing balm upon my wounded heart—

Where sorrow dwelt, now joy begins to reign,

Your lips the medicine that makes me whole again.

What sleepless nights and aching days had torn,

One tender kiss has perfectly restored.

Sweet remedy pressed soft against my mouth,

You kissed away my winter, brought me south.

Explanatory Note: Medela is a rhetorical figure that presents something as a cure or remedy, often used metaphorically to describe how one thing heals or resolves another. In this poem, the speaker employs medela to transform a kiss into medicinal imagery—presenting the beloved's kiss as a healing balm, cure, and remedy that restores emotional and spiritual wholeness, elevating the romantic gesture to the level of miraculous healing power.

Kiss Bliss (with Mycterismus)

Oh yes, that kiss was surely worth the wait—

How masterfully you pursed your lips like stone!

Such passion in your cold, reluctant state,

The way you turned your cheek and pulled away.

Truly, no lover could ask for more delight

Than feeling wanted with such clear disdain,

Your kiss as warm as winter's bitter night,

As tender as a blade across the vein.

Explanatory Note: Mycterismus is a rhetorical figure of scornful mockery or sneering contempt, often expressed through bitter irony and sarcastic praise. In this poem, the speaker employs mycterismus to mock a disappointing romantic encounter, using ironic language that praises what is actually being criticized—the coldness, reluctance, and lack of genuine affection in the kiss, transforming romantic disappointment into cutting satirical commentary.

Kiss Bliss (with Sarcasmus)

Oh wonderful, another stolen kiss

From lips that promise love but speak of lies—

How thrilling to taste such false bliss,

To feel betrayal dressed in lover's guise.

Yes, kiss me with that practiced, hollow art,

Show me again how well you break a heart.

Such tender cruelty, such sweet deceit,

Your kiss makes my destruction feel complete.

Explanatory Note: Sarcasmus is a rhetorical figure involving bitter, cutting irony that expresses contempt or scorn through seemingly positive language that actually conveys the opposite meaning. In this poem, the speaker employs sarcasmus to express pain and disillusionment about a deceptive kiss, using words like "wonderful," "thrilling," and "tender" in a bitterly ironic way that reveals the true depth of hurt and betrayal beneath the surface praise.

Kiss Bliss (with Epiplexis)

How could you kiss me with such sweet delight,

Then turn away as if it meant nothing?

What kind of heart can burn with passion's light,

Yet speak of love like it were just pretending?

Why do you wound me with your tender touch,

Then act as though my feelings matter not?

Is this how lovers treat what means so much—

With kisses given, then so soon forgot?

Explanatory Note: Epiplexis is a rhetorical figure involving emotional rebuke or reproachful questioning, typically expressed through a series of accusatory or bewildered questions directed at someone who has caused hurt or disappointment. In this poem, the speaker employs epiplexis to confront their lover about the contradiction between an intimate kiss and subsequent coldness, using reproachful questions to express confusion, hurt, and indignation at being treated with such apparent inconsistency.

Kiss Bliss (with Percontatio)

Tell me, what power lies within a kiss

That it can stop the turning of the earth?

Who taught your lips such magic as this,

To weave a spell that measures more than worth?

Can mortal touch contain such divine fire?

Do angels whisper secrets through your breath?

What ancient gods conspired to inspire

This kiss that conquers even fear of death?

Explanatory Note: Percontatio is a rhetorical figure involving earnest inquiry or questioning for the purpose of investigation, typically used to explore the nature or cause of something through a series of genuine, wondering questions. In this poem, the speaker employs percontatio to investigate the mysterious power of a kiss, using sincere questions to explore its almost supernatural effects and transformative nature, treating the romantic moment as a phenomenon worthy of philosophical inquiry.

Kiss Bliss (with Categoria)

You are a thief who steals with gentle lips,

A liar speaking truth through tender touch—

Your kiss commits the crime of sweet eclipse,

Erasing all the world with passion's clutch.

I name you guilty of this heart's surrender,

Accuse you of conspiracy with night,

Charge you with making moments burn like tender,

And sentence you to love me by first light.

Explanatory Note: Categoria is a rhetorical figure involving direct accusation or categorical denunciation, often used to formally charge someone with specific offenses or wrongdoings. In this poem, the speaker employs categoria to playfully "accuse" their lover of various romantic "crimes"—theft, conspiracy, and heart-stealing—using legal language to transform the experience of being kissed into a mock trial where love itself becomes the delightful offense for which the beloved stands charged.

Kiss Bliss (with Proclees)

Come then, if you dare claim your lips are wise,

And prove your boasted skill in love's sweet art—

Show me how passion burns behind your eyes,

If you possess the power to stop my heart.

Challenge me with that kiss you say is true,

Display the tenderness you think I lack,

I'll gladly let you teach me what you knew,

If you can make me never want you back.

Explanatory Note: Proclees is a rhetorical figure that emphasizes a point by framing it positively while challenging one's adversary. In this poem, the speaker employs proclees to issue a romantic challenge to their lover, daring them to prove their romantic prowess through their kiss—using positive framing ("if you dare," "show me," "prove") to create an enticing provocation that simultaneously challenges and invites intimate connection.

Kiss Bliss (with Bdelygmia)

That kiss—oh vile betrayer of my peace!

Disgusting trap that snared my willing heart,

Repulsive sweetness that will never cease

To haunt me with its loathsome, tender art.

Abhorrent lips that spoke of love so true,

Detestable the way you made me yearn—

I curse the moment that I tasted you,

Yet pray for one more kiss for which I burn.

Explanatory Note: Bdelygmia is a rhetorical figure expressing disgust, loathing, or abhorrence, often using language of revulsion to describe something the speaker finds repugnant. In this poem, the speaker employs bdelygmia to express conflicted feelings about a kiss that has left them emotionally devastated—using words like "vile," "disgusting," "repulsive," and "abhorrent" to convey their simultaneous revulsion and attraction, creating a paradox where the very thing they claim to detest is also what they desperately desire.

Kiss Bliss (with Cataplexis)

Your kiss struck down my senses like a blade,

Left me astonished, breathless, overthrown—

All reason fled, all careful guards unmade,

Stunned by the power in your lips alone.

I stood there dumbstruck by love's sudden force,

Bewildered by the lightning in your touch,

Amazed to find my heart knocked from its course

By one small kiss that conquered oh so much.

Explanatory Note: Cataplexis is a rhetorical figure expressing amazement, astonishment, or being struck with wonder—often describing a state of being overwhelmed or stunned by something unexpected or powerful. In this poem, the speaker employs cataplexis to convey the overwhelming impact of a kiss that leaves them completely astonished and disoriented, using language of shock and wonder ("struck down," "astonished," "stunned," "dumbstruck") to capture the transformative power of unexpected romantic intensity.

Kiss Bliss (with Ara)

May all the gods curse that deceiving kiss

That promised heaven but delivered hell—

Let sorrow haunt the lips that brought me this

False bliss that broke my heart's protective shell.

May bitter winds forever dry the mouth

That spoke of love then turned away in scorn,

And may all joy abandon you, uncouth

Betrayer, leaving you forever torn.

Explanatory Note: Ara is a rhetorical figure involving a curse, malediction, or imprecation—calling down divine punishment or ill fortune upon someone or something. In this poem, the speaker employs ara to curse the kiss and the lips that delivered romantic betrayal, using formal cursing language ("May all the gods curse," "Let sorrow haunt," "May bitter winds") to transform hurt and disappointment into ritualized condemnation, elevating personal romantic grievance into the realm of divine justice.

Kiss Bliss (with Cursing)

Damn that kiss that set my soul on fire,

Hell take the lips that made me feel so much—

Curse the moment passion raised desire,

And blast the memory of your tender touch.

To hell with all the sweetness that you gave,

Damn every whisper, every stolen breath,

May devils take the love I couldn't save,

And curse this heart that loves you unto death.

Explanatory Note: Cursing is a rhetorical figure involving the use of profane or blasphemous language to express strong emotion, typically anger, frustration, or passionate intensity. In this poem, the speaker employs cursing to express the overwhelming and conflicted emotions surrounding a powerful romantic encounter, using profane language ("damn," "hell," "blast," "devils") to convey both the intensity of their passion and their frustration with love's destructive power, creating a paradox where they curse what they simultaneously cherish.

Kiss Bliss (with Eulogia)

Blessed be the lips that touched mine in the night,

Praise to the mouth that spoke love's sacred word—

Holy the kiss that made my darkness bright,

Divine the whisper that my heart overheard.

Blessed be the moment when our breaths became one,
Sacred the touch that made my spirit whole,

Praise to the love that outshines moon and sun,

Holy the kiss that blessed my weary soul.

Explanatory Note: Eulogia is a rhetorical figure involving blessing, praise, or speaking well of someone or something, often using sacred or reverent language to express deep appreciation and honor. In this poem, the speaker employs eulogia to transform a romantic kiss into a sacred experience, using blessing language ("Blessed be," "Praise to," "Holy," "Divine," "Sacred") to elevate the intimate moment into something approaching religious reverence, treating love as a form of spiritual blessing.

Kiss Bliss (with Paeanismus)

Victory! Your kiss has conquered all my doubt,

Triumph rings through every beating vein—

I sing of love that puts all fear to rout,

And celebrate this sweet, victorious pain.

Hail to the lips that won my willing heart,

Glory to passion's overwhelming might!

Let all the world rejoice at love's great art,

Your kiss brings dawn to end my lonely night

Explanatory Note: Paeanismus is a rhetorical figure involving triumphant celebration or exuberant praise, derived from the ancient Greek paean (a song of triumph or praise). In this poem, the speaker employs paeanismus to celebrate a kiss as a glorious victory, using language of triumph and celebration ("Victory!" "Triumph," "Hail," "Glory," "rejoice") to transform the romantic moment into a cause for jubilant proclamation, treating love's conquest as worthy of public celebration and praise.

Kiss Bliss (with Comprobatio)

Your kiss, I must confess, surpassed all dreams—
Indeed, no poet's verse could capture quite

The way your lips made starlight pale, it seems,

And turned my darkness into blazing light.

Truly, I cannot deny the power held

Within that single, tender, sacred touch—

Yes, every defense my heart once upheld

Was conquered by a kiss that meant so much

Explanatory Note: Comprobatio is a rhetorical figure involving confirmation, approval, or acknowledgment of something as true or valid, often expressed through phrases like "indeed," "truly," "I must confess," or "I cannot deny." In this poem, the speaker employs comprobatio to validate and confirm the extraordinary power of a kiss, using language of acknowledgment and testimony to give credence to love's transformative effects, treating the romantic experience as an undeniable truth that demands recognition.

Kiss Bliss (with Parrhesia)

I'll speak the truth though it may wound us both—

Your kiss means nothing if your heart's not there,

And I would rather bear love's painful oath

Than taste false sweetness poisoned by despair.

Let me be blunt: I'd take one honest word

Over a thousand kisses born of lies,

For though my frankness leaves us both unstirred,

At least we'll know what's real behind our eyes.

Explanatory Note: Parrhesia is a rhetorical figure involving fearless, frank speech or bold truth-telling, often used when speaking candidly about difficult or potentially offensive matters. In this poem, the speaker employs parrhesia to address the painful reality of a potentially hollow kiss, choosing brutal honesty over comfortable illusion. The figure allows the speaker to confront uncomfortable truths about love and authenticity, valuing sincere communication over pleasant deception.

Kiss Bliss (with Eucharistia)

I thank the fates that brought your lips to mine,

Grateful for every breath we shared as one—

How blessed I am to taste love so divine,

And witness passion's miracle begun.

My heart gives thanks for tender mercies shown

In kisses sweeter than the morning dew,

Thankful that I need never love alone,

Forever grateful that I found in you.

Explanatory Note: Eucharistia is a rhetorical figure expressing gratitude, thanksgiving, or appreciation, often used to acknowledge blessings received or favors granted. In this poem, the speaker employs eucharistia to express profound gratitude for the gift of romantic love, using language of thanksgiving ("I thank," "grateful," "blessed," "gives thanks," "thankful") to transform the experience of being kissed into an occasion for appreciative acknowledgment of love as a divine blessing worthy of heartfelt gratitude.

Kiss Bliss (with Syngnome)

Forgive me if I trembled at your touch,

Pardon my breathlessness when lips met mine—

I know it seems I felt it all too much,

But who could help but worship at love's shrine?

Excuse my heart for beating far too fast,

And understand if words became too few,

For when such perfect moments come to pass,

What mortal soul could help but fall to you?

Explanatory Note: Syngnome is a rhetorical figure expressing pardon, forgiveness, or understanding, often used to excuse one's behavior or reactions by appealing to common human nature or circumstances. In this poem, the speaker employs syngnome to seek understanding for their overwhelming emotional response to a kiss, asking pardon for trembling, breathlessness, and being overcome, while suggesting that such reactions are natural and forgivable given the power of the romantic moment.

Kiss Bliss (with Exclamatio)

Oh! What fire burns within that sacred kiss,

Ah! How it sets my very soul aflame—

Behold! The power held in tender bliss,

Alas! I'll never be quite the same. How sweet!

The way your lips found mine at last, What joy!

To feel love's overwhelming might, Oh rapture!

Time itself forgotten fast, What bliss!

To kiss you in the fading light.

Explanatory Note: Exclamatio is a rhetorical figure involving passionate outcry or emotional exclamation, typically expressed through interjections like "Oh!" "Ah!" "Behold!" "Alas!" "How!" and "What!" In this poem, the speaker employs exclamatio to express the intense emotions aroused by a kiss, using a series of exclamatory phrases to convey overwhelming passion, wonder, and joy, transforming the romantic moment into a cascade of emotional outbursts that capture the speaker's inability to contain their feelings.

Kiss Bliss (with Exuscitatio)

Wake up, my heart, and feel this tender fire!

Arise, my soul, from slumber's darkest deep—

Stir now, all dormant passion and desire,

No longer let love's promise silent sleep.

Awaken, dreams that long have lain in wait,

Come forth, sweet hopes that trembled shy and still,

Rise up and claim your blessed, kissed fate,

Let love's first kiss your destiny fulfill.

Explanatory Note: Exuscitatio is a rhetorical figure involving arousal or awakening, used to stir up emotions, passions, or dormant feelings, often through direct address to the heart, soul, or other aspects of the self. In this poem, the speaker employs exuscitatio to awaken their own dormant romantic feelings in response to a transformative kiss, using imperative language ("Wake up," "Arise," "Stir," "Awaken," "Rise up") to rouse all aspects of their being to embrace love's awakening power.

Kiss Bliss (with Paralipsis)

I will not speak of how your kiss undid

My careful walls with one sweet, tender touch—

Nor mention how my racing heart was bid

To love you more than I should dare too much.

I shall not tell how breathless I became,

Or how your lips left mine forever changed,

I won't confess this kiss set me aflame—

These truths are far too sacred to be ranged.

Explanatory Note: Paralipsis (also called praeteritio) is a rhetorical figure in which the speaker mentions something by claiming not to mention it, or brings attention to a topic while pretending to pass over it. In this poem, the speaker employs paralipsis to reveal the profound effects of a kiss while ostensibly refusing to discuss them, using phrases like "I will not speak," "Nor mention," "I shall not tell," and "I won't confess" to actually emphasize and highlight the very romantic experiences they claim to be avoiding.

Made in United States
Troutdale, OR
11/03/2026

56845126R00086